Springer Series in Optical Sciences

Volume 218

Springer Series in Optical Sciences is led by Editor-in-Chief William T. Rhodes, Georgia Institute of Technology, USA, and provides an expanding selection of research monographs in all major areas of optics:

- lasers and quantum optics
- ultrafast phenomena
- optical spectroscopy techniques
- optoelectronics
- information optics
- applied laser technology
- industrial applications and
- other topics of contemporary interest.

With this broad coverage of topics the series is useful to research scientists and engineers who need up-to-date reference books.

More information about this series at http://www.springer.com/series/624

Paulo A. Ribeiro · Maria Raposo
Editors

Optics, Photonics and Laser Technology

 Springer

Editors
Paulo A. Ribeiro
Faculdade de Ciências e Tecnologia
Universidade Nova de Lisboa
Caparica, Portugal

Maria Raposo
Faculdade de Ciências e Tecnologia
Universidade Nova de Lisboa
Caparica, Portugal

ISSN 0342-4111   ISSN 1556-1534   (electronic)
Springer Series in Optical Sciences
ISBN 978-3-030-07506-4   ISBN 978-3-319-98548-0   (eBook)
https://doi.org/10.1007/978-3-319-98548-0

This Springer imprint is published by the registered company Springer Nature Switzerland AG
The registered company address is: Gewerbestrasse 11, 6330 Cham, Switzerland

Preface

The present book includes reviews on selected relevant scientific themes inspired on papers presented at the 4th International Conference on Photonics, Optics and Laser Technology (PHOTOPTICS 2016), held in Rome, Italy, from 27 to 29 February 2016. The event received 98 paper submissions from 29 countries, which have been selected by the chairs, based on a set criteria items that include the assessment and comments provided by the program committee members, the session chairs' assessment and also the program chairs' global view of all contributions selected to be included in the technical program. Under this compliance, this book covers both theoretical and practical aspects within Optics, Photonics and Laser Technologies' main tracks, with the chapter contents contributing to the understanding of current research trends. The book contents, spread through eleven chapters, cover imaging and microscopy—techniques and apparatus, with contributions as the status quo of waveguide evanescent fluorescence and scattering microscopy, micro-lenslet arrays' characterization through holographic interferometer microscopy, performance of segmented reflecting telescopes; optical communications where estimation of the OSNR penalty as a result of in-band crosstalk is outlined; laser safety, where high accuracy laser pointer assessment is addressed; biomedical developments and applications including contributions as mitigation of speckle noise in optical coherence tomograms toward dynamics of blood perfusion and tissue oxygenation, tissue's conservation assessment via coherent hemodynamic spectroscopy and, finally, the field of photonic materials and devices gathers the contributions, CMOS-based photomultiplier devices, photochromic materials toward energy harvesting and novel photo-thermo-refractive for photonic, plasmonic, fluidic and luminescent devices.

We thank all the authors for their contributions and also to the reviewers who have helped ensuring the quality of this publication.

Caparica, Portugal

Maria Raposo
Paulo A. Ribeiro

Organization

Conference Chair

Paulo A. Ribeiro, CEFITEC/FCT/UNL, Portugal

Program Chair

Maria Raposo, CEFITEC, FCT/UNL, Portugal

Program Committee

Jean–Luc Adam, Université de Rennes 1—CNRS, France
Kamal E. Alameh, Edith Cowan University, Australia
Marcio Alencar, Universidade Federal de Sergipe, Brazil
Augusto A. Iribarren Alfonso, Instituto de Ciencia y Tecnología de Materiales, Universidad de La Habana, Cuba
Tatiana Alieva, Universidad Complutense de Madrid, Spain
Hakan Altan, Terahertz Research Laboratory, Physics Department, METU, Turkey
António Amorim, Departamento de Física da Faculdade de Ciências da Universidade de Lisboa, Portugal
Alexander Argyros, Institute of Photonics and Optical Science, University of Sydney, Australia
Andrea Armani, University of Southern California, USA
Jhon Fredy Martinez Avila, Universidade Federal de Sergipe, Brazil
Morten Bache, Technical University of Denmark, Denmark
Vanderlei Salvador Bagnato, Instituto de Física de São Carlos/Universidade de São Paulo, Brazil
Francesco Baldini, Istituto di Fisica Applicata "Nello Carrara," Italy
Pedro Barquinha, Departamento de Ciência dos Materiais, Faculdade de Ciências e Tecnologia, Universidade Nova de Lisboa, Portugal
Almut Beige, University of Leeds, UK
Alessandro Bettini, Università di Padova, Italy

Robert Ferguson, National Physical Laboratory, UK
Paulo Torrão Fiadeiro, Universidade da Beira Interior, Portugal
José Figueiredo, Universidade do Algarve, Portugal
Luca Fiorani, Researcher at ENEA—Professor at "Lumsa," "Roma Tre," and "Tor Vergata" Universities, Italy
Orlando Frazão, INESC Porto, Portugal
Alberto Garcia, Instituto de Física dos Materiais da Universidade do Porto, Spain
Ivana Gasulla, Polytechnic University of Valencia, Spain
Sanka Gateva, Institute of Electronics, Bulgaria
Malte Gather, University of St. Andrews, UK
Marco Gianinetto, Politecnico di Milano, Italy
John Girkin, Department of Physics, Biophysical Sciences Institute, Durham University, UK
Guillaume Gomard, Light Technology Institute, Germany
Yandong Gong, Institute for Infocomm Research, Singapore
Giuseppe Grosso, Universita' di Pisa, Italy
Norbert Grote, Fraunhofer Heinrich Hertz Institute, Germany
Mircea D. Guina, Tampere University of Technology, Finland
Young-Geun Han, Hanyang University, Korea, Republic of
Joseph W. Haus, University of Dayton, USA
Rosa Ana Perez Herrera, Universidad Publica de Navarra, Spain
Daniel Hill, Universitat de Valencia, Spain
Peter Horak, University of Southampton, UK
Iwao Hosako, National Institute of Information and Communications Technology, Japan
Silvia Soria Huguet, IFAC-N. Carrara Institute of Applied Physics/C.N.R., Italy
Nicolae Hurduc, Gheorghe Asachi Technical University of Iasi, Romania
Baldemar Ibarra-Escamilla, Instituto Nacional de Astrofísica, Optica y Electrónica, Mexico
Jichai Jeong, Korea University, Korea, Republic of
Phil Jones, Optical Tweezers Group, Department of Physics and Astronomy, University College London, UK
José Joatan Rodrigues Jr., Universidade Federal de Sergipe, Brazil
Yoshiaki Kanamori, Tohoku University, Japan
Fumihiko Kannari, Keio University, Japan
Pentti Karioja, VTT Technical Research Centre of Finland, Finland
Christian Karnutsch, Karlsruhe University of Applied Sciences, Faculty of Electrical Engineering and Information Technology, Germany
Miroslaw A. Karpierz, Faculty of Physics, Warsaw University of Technology, Poland
Henryk Kasprzak, Wroclaw University of Technology, Poland
Chulhong Kim, POSTECH, Republic of Korea
Young-Jin Kim, Nanyang Technological University, Singapore
Alper Kiraz, Koç University, Turkey
Stefan Kirstein, Humboldt-Universitaet zu Berlin, Germany

Gang-Ding Peng, University of New South Wales, Australia
Alfons Penzkofer, Universität Regensburg, Germany
Luís Pereira, Faculdade de Ciências e Tecnologia, Universidade Nova de Lisboa, Portugal
Romain Peretti, Swiss Federal Institute of Technology in Zurich (ETHZ), Switzerland
Francisco Pérez-Ocón, Universidad de Granada, Spain
Klaus Petermann, Technische Universität Berlin, Germany
Angela Piegari, ENEA, Italy
Armando Pinto, Instituto de Telecomunicações—University of Aveiro, Portugal
João Pinto, Aveiro University, Portugal
Sandro Rao, University Mediterranea of Reggio Calabria, Italy
Maria Raposo, CEFITEC, FCT/UNL, Portugal
Alessandro Restelli, Joint Quantum Institute, USA
Paulo A. Ribeiro, CEFITEC/FCT/UNL, Portugal
Mauricio Rico, Centro de Láseres de Pulsados (CLPU), Spain
Giancarlo C. Righini, Enrico Fermi Centre & IFAC CNR, Italy
Alessandro Rizzi, Università di Milano, Italy
Murilo Romero, University of Sao Paulo, Brazil
Luis Roso, Centro de Laseres Pulsados, CLPU, Spain
Luigi L. Rovati, Università di Modena e Reggio Emilia, Italy
Svilen Sabchevski, Institute of Electronics, Bulgarian Academy of Sciences, Bulgaria
Manuel Pereira dos Santos, Centro de Física e Investigação Tecnológica, Faculdade de Ciências e Tecnologia, Universidade Nova de Lisboa, Portugal
Christian Seassal, CNRS, University of Lyon, France
Wei Shi, Tianjin University, China
Ronaldo Silva, Federal University of Sergipe, Brazil
Susana Silva, INESC TEC, Portugal
Feng Song, Nankai University, China
Viorica Stancalie, National Institute for Laser, Plasma and Radiation Physics, Romania
Slawomir Sujecki, University of Nottingham, UK
Xiao Wei Sun, Nanyang Technological University, Singapore
João Manuel R. S. Tavares, FEUP—Faculdade de Engenharia da Universidade do Porto, Portugal
Ladislav Tichy, IMC ASCR, Czech Republic
Swee Chuan Tjin, Nanyang Technological University, Singapore
José Roberto Tozoni, Universidade Federal de Uberlandia, Brazil
Din-Ping Tsai, National Taiwan University, Taiwan
Kevin Tsia, The University of Hong Kong, Hong Kong
Graham A. Turnbull, University of St. Andrews, UK
Eszter Udvary, Optical and Microwave Telecommunication Laboratory, Budapest University of Technology and Economics (BME), Hungary
M. Selim Ünlü, Boston University, USA

Carlos M. S. Vicente, Instituto de Telecomunicações and CICECO—Aveiro, Portugal
Wanjun Wang, Louisiana State University, USA
Zinan Wang, University of Electronic Science and Technology, China
James Wilkinson, Faculty of Physical and Applied Sciences University of Southampton, UK
Lech Wosinski, KTH—Royal Institute of Technology, Sweden
Jingshown Wu, Networked Communications Program (NCP), Taiwan
Kaikai Xu, University of Electronic Science and Technology of China (UESTC), China
Minghong Yang, Wuhan University of Technology, China
Anna Zawadzka, Institute of Physics, Faculty of Physics, Astronomy and Informatics Nicolaus Copernicus University, Poland
He-Ping Zeng, East China Normal University, China
Lin Zhang, Aston University, UK

Additional Reviewers

Alex Araujo, Instituto Federal de Mato Grosso do Sul, Brazil

Invited Speakers

Giancarlo C. Righini, Enrico Fermi Centre & IFAC CNR, Italy
Andrea Cusano, University of Sannio, Italy
Manuel Filipe Costa, Universidade do Minho, Portugal

Contents

Contributors

D. Abi Haidar IMNC Laboratory, UMR 8165, CNRS/IN2P3, Paris-Saclay University, Orsay, Paris, France; Paris Diderot University, Paris, France

Saba Adabi Department of Biomedical Engineering, Wayne State University, Detroit, USA; Department of Applied Electronics, Roma Tre University, Rome, Italy

V. Aseev Department of Optical Information Technologies and Materials, ITMO University, Saint Petersburg, Russia

Adolfo V. T. Cartaxo Optical Communications and Photonics, Instituto de Telecomunicações, Lisbon, Portugal; Instituto Universitário de Lisboa (ISCTE-IUL), Lisbon, Portugal

Anne Clayton Department of Biomedical Engineering, Wayne State University, Detroit, USA

Silvia Conforto Department of Applied Electronics, Roma Tre University, Rome, Italy

N. D'Ascenzo Huazhong University of Science and Technology, Wuhan, China

B. Devaux Department of Neurosurgery, Sainte Anne Hospital, Paris, France

V. Dubrovin Department of Optical Information Technologies and Materials, ITMO University, Saint Petersburg, Russia

Sergio Fantini Department of Biomedical Engineering, Tufts University, Medford, MA, USA

Pedro Farinha CEFITEC, Departamento de Física, Faculdade de Ciências e Tecnologia, Universidade Nova de Lisboa, Caparica, Portugal

Ali Hojjat School of Physical Sciences, University of Kent, Canterbury, UK

A. Ibrahim IMNC Laboratory, UMR 8165, CNRS/IN2P3, Paris-Saclay University, Orsay, Paris, France

A. Ignatiev Department of Optical Information Technologies and Materials, ITMO University, Saint Petersburg, Russia

S. Ivanov Department of Optical Information Technologies and Materials, ITMO University, Saint Petersburg, Russia

Varun Kumar Laser Applications and Holography Laboratory, Instrument Design Development Centre, Indian Institute of Technology Delhi, New Delhi, India

Gonçalo Magalhães-Mota CEFITEC, Departamento de Física, Faculdade de Ciências e Tecnologia, Universidade Nova de Lisboa, Caparica, Portugal

Silvia Mittler Department of Physics and Astronomy, The University of Western Ontario, London, ON, Canada

Mohammadreza Nasiriavanaki Department of Biomedical Engineering, Wayne State University, Detroit, USA

N. Nikonorov Department of Optical Information Technologies and Materials, ITMO University, Saint Petersburg, Russia

Kanokwan Nontapot National Institute of Metrology, Pathum Thani, Thailand

Bruno R. Pinheiro Optical Communications and Photonics, Instituto de Telecomunicações, Lisbon, Portugal; Instituto Universitário de Lisboa (ISCTE-IUL), Lisbon, Portugal

Adrian G. Podoleanu Applied Optics Group, University of Kent, Canterbury, UK

F. Poulon IMNC Laboratory, UMR 8165, CNRS/IN2P3, Paris-Saclay University, Orsay, Paris, France

Narat Rujirat National Institute of Metrology, Pathum Thani, Thailand

Maria Raposo CEFITEC, Departamento de Física, Faculdade de Ciências e Tecnologia, Universidade Nova de Lisboa, Caparica, Portugal

João L. Rebola Optical Communications and Photonics, Instituto de Telecomunicações, Lisbon, Portugal; Instituto Universitário de Lisboa (ISCTE-IUL), Lisbon, Portugal

Paulo A. Ribeiro CEFITEC, Departamento de Física, Faculdade de Ciências e Tecnologia, Universidade Nova de Lisboa, Caparica, Portugal

Angelo Sassaroli Department of Biomedical Engineering, Tufts University, Medford, MA, USA

V. Saveliev Huazhong University of Science and Technology, Wuhan, China

Susana Sério CEFITEC, Departamento de Física, Faculdade de Ciências e Tecnologia, Universidade Nova de Lisboa, Caparica, Portugal

Y. Sgibnev Department of Optical Information Technologies and Materials, ITMO University, Saint Petersburg, Russia

Chandra Shakher Laser Applications and Holography Laboratory, Instrument Design Development Centre, Indian Institute of Technology Delhi, New Delhi, India

A. Sidorov Department of Optical Information Technologies and Materials, ITMO University, Saint Petersburg, Russia

Maricor Soriano National Institute of Physics, University of the Philippines, Diliman, Quezon City, Philippines

Giovanni Tapang National Institute of Physics, University of the Philippines, Diliman, Quezon City, Philippines

Kristen T. Tgavalekos Department of Biomedical Engineering, Tufts University, Medford, MA, USA

P. Varlet Department of Neuropathology, Sainte Anne Hospital, Paris, France

Q. Xie Huazhong University of Science and Technology, Wuhan, China

M. Zanello IMNC Laboratory, UMR 8165, CNRS/IN2P3, Paris-Saclay University, Orsay, Paris, France; Department of Neurosurgery, Sainte Anne Hospital, Paris, France

Xuan Zang Department of Biomedical Engineering, Tufts University, Medford, MA, USA

Chapter 1
Waveguide Evanescent Field Fluorescence and Scattering Microscopy: The Status Quo

Silvia Mittler

Abstract In the last few years Waveguide Evanescent Field Fluorescence (WEFF) and Scattering (WEFS) microscopy were developed which are alternatives to TIR and TIRF microscopy. Both technologies implement a slab waveguide-microscopy chip with a coupling grating. The technologies are described and compared to TIR and TIRF microscopy. The advantages of the waveguide method are clearly addressed. A brief history of the technology's development and similar activities in the field are discussed. Application examples from both WEF microscopies follow: static distance mapping with a multimode waveguide, dynamic solubilisation studies of cell plasma membranes and the kinetic response of osteoblasts to trypsin (WEFF); bacteria sterilization as well as cell adhesion and granularity studies (WEFS). The combination of both methods is discussed and found not suitable. In order to mass fabricate the necessary waveguide chips with the grating an all-polymer-waveguide chip was developed. This should allow to bring the new microscopy methods to the interested scientific community.

1.1 Introduction

With the aim of developing new medical devices with direct tissue contact, drug delivery vehicles, and tissue engineering scaffolds, there has been increasing interest in recent years in the interactions of cells with both synthetic and natural biomaterials [1, 2]. In particular, the study of the contact regions between a cell and its substratum is of considerable interest as its investigation delivers *inter alia* information about the cytocompatibility of the substratum—the affinity of cells towards that particular surface. Promotion or inhibition of cell adhesion to synthetic and natural biomaterials is often crucial to the proper function of a particular device. Some information concerning these interactions, e.g. the lateral location and the density of

S. Mittler (✉)
Department of Physics and Astronomy, The University
of Western Ontario, London, ON, Canada
e-mail: smittler@uwo.ca

© Springer Nature Switzerland AG 2018
P. A. Ribeiro and M. Raposo (eds.), *Optics, Photonics and Laser Technology*, Springer
Series in Optical Sciences 218, https://doi.org/10.1007/978-3-319-98548-0_1

the adhesion sites, as well as their relationship to the actin stress fiber system, part of the cell's cytoskeleton, can be inferred from fluorescence microscopy of immuno-labeled molecules involved in adhesion; typically, vinculin, a protein located within the multi-protein complex that anchors the adhesion to the cytoskeleton inside the cell [3]. These methods only deliver signals from the focus volume and no information about adhesion distances to the substratum. However, a direct and quantitative method to address the distance to the substratum is highly attractive. To address this need, different microscopic techniques based on electron microscopy [4] and optical means such as evanescent fields and interference techniques have been developed. Total internal reflection fluorescence (TIRF) [3, 5], surface plasmon resonance microscopy (SPRM) [6], interference fluorescence microscopy (IRM) [7], fluorescence interference contrast (FLIC) microscopy [8] and combinations thereof [3, 9] have been used to visualize and quantify these contacts. The contacts themselves had been discovered by interference reflection microscopy (IRM) in the 1970s [10].

Bacteria, on the other hand, are the most metabolically diverse group of organisms found in all natural environments including air, water and soil. Bacteria commonly occur with food sources and are also found within and on our bodies. However, concerns exist over contamination of food, water, and air by pathogenic bacteria [11] that can enter our bodies through ingestion, inhalation, cuts or lacerations [12]. Therefore, there is an increasing interest in bacterial contamination and the need for anti-bacterial surfaces not only for application in the food industry but also for medical and hygienic purposes [13]. Over two million hospital-acquired cases of infection are reported annually in the USA, which lead to approximately 100,000 deaths annually and added nearly $5 billion to U.S. healthcare costs [14, 15]. Contamination of medical devices (e.g., catheters and implants) has been attributed to 45% of these infections [16]. Bacterial contamination of any surface typically begins with the initial adhesion of only a few cells that can then develop into a more structurally cohesive biofilm in less than 24 h when provided with suitable nutrient conditions sustaining metabolism and cell division [17]. Therefore, a better understanding of bacterial adhesion to surfaces is important for technical surface development and in biomedical applications. However, the precise measurement of bacterial adhesion to surfaces are difficult and time consuming because bacterial cells typically occur on the micrometer-scale and their adhesion forces are generally low, typically 0.1–100 nN [18]. Recent studies on the detection of bacteria on surfaces have focused on similar imaging systems as with cells such as optical [19] and fluorescent microscopy [20] to image the bacteria themselves or luminescence measurement of the presence of cells by ATP (adenosine triphosphate) detection systems [21] Surface Plasmon Resonance (SPR) sensors [22], Nucleic Acid Detection [23], Optical Waveguide Lightmode Spectroscopy (short: waveguide spectroscopy) [24], Optical Leaky Waveguide Sensors [25], and Evanescent Mode Fiber Optic Sensors [26] have also been applied in order to detect biochemical toxins as signatures of bacteria.

In conclusion, it is important to have methods available which are able to investigate interfaces between a technical surface and a bacterium or cell.

In recent years, Total Internal Reflection Fluorescent (TIRF) microscopy has been demonstrated to be an effective method for studying cell-substrate interactions that occur at surfaces and interfaces. Using TIRF microscopy, the behavior of various types of cells [27, 28] and bacteria [29, 30] near surfaces has been characterized. Total Internal Reflection (TIR) Microscopy utilizes the basic technology of TIRF without any fluorescence dyes present in the sample by creating an optical contrast due to scattering [31]. Recent studies have also demonstrated the use of TIR for imaging microbial adhesion.

This paper will give a brief history and literature overview as well as a review on biophysical applications of WEFF and WEFS microscopy on cells and bacteria and a short outlook on current developments on mass producible all-polymer-waveguide-chips to offer the methods to a broader user base.

1.2   Brief History

The waveguide evanescent field scattering (WEFS) technique was developed by Thoma et al. in 1997 [32, 33] for ultrathin technical structures on surfaces using conventional ion exchanged waveguides. Thoma et al. investigated the influence of the polarization direction of the mode and the mode number on the achieved scattering image contrast. It was found that with increasing mode number the contrast increased. TE modes depicted a better contrast than TM modes. Immersion of the samples in an aqueous solution decreased the contrast as the refractive index difference between scattering centers and the surrounding medium decreased.

Before the first cells were imaged evanescently with a waveguide, a series of approaches were made to combine established microscopy methods with a conventional waveguide and waveguide spectroscopy to on-line monitor e.g. adhesion and proliferation of cultured cells [34–36] These attempts suffered from the small penetration depth of the evanescent fields into the cells of the used conventional waveguides. Therefore, Horvath et al. [37] proposed in 2005 reverse symmetry waveguides which provided deeper penetrating evanescent fields in these combination technologies.

The first waveguide evanescent field fluorescence (WEFF) experiment with commercial mono-mode glass waveguides (designed for sensing on coupling gratings) on cells was shown by Grandin et al. [38]. Vinculin staining was carried out on fixed human fibroblast cells. The coupling was carried out via a coupling grating located within the sample area. The image suffered from substantial artefacts and was very noisy in form of stripes. The grating position in the field of view, lead to too many scattered photons and to resonantly out-coupled light.

WEFF microscopy with ion exchanged mono- or multi-mode glass waveguides and stained cell's plasma membranes was developed by Hassanzadeh et al. [39] in 2008 as a straightforward alternative to TIRF microscopy for imaging ultrathin films and cell-substrate interaction. The mode coupling was achieved via a grating coupler outside the imaging area. The image quality was increasing, but artefacts due to scattered light were still seen in the images and are not completely avaoidable. TIRF images share the same issues [40]. Hassanzadeh et al. have then shown applications of WEFF microscopy with static and dynamic investigations [39, 41, 42]. They were the first who used a multi-mode waveguides with mode selective grating coupling to determine distances of cell adhesions to the waveguide surface [43].

A year later, 2009, Agnarsson et al. [44] presented a symmetric waveguide structure for WEFF microscopy where the cladding material is index-matched to the sample solution (aqueous media). The optical chips were fabricated from polymers involving standard cleanroom technologies such as spin-coating, photolithography and dry etching (core: polymethylmethacrylate (PMMA) and substrate: amorphous perfluorinated optical polymer (Cytop)). Agnarsson et al. coupled via end-fire coupling mode-insensitive into a mono-mode waveguide. MCF7 breast cancer cells immune-stained with monoclonal antibody against the transmembrane adhesion protein E-cadherin (HECD-1) and Alexa Fluor 546 Goat Anti Mouse IgG1 fluorescent secondary antibody. Very clear WEFF images were obtained with some minor scattering artefacts. Agnarsson et al. were combining WEFF microscopy on the mono-mode polymer symmetric waveguide structures with waveguide spectroscopy and channel waveguide operation (directional couplers, ring resonators, Mach-Zehnder interferometers, etc.) for sensing application and on-chip control of illumination with sub-millisecond control [45–48].

In 2014, Nahar et al. [49] implemented WEFS microscopy for bacteria studies and started to investigate cultured ostoeblasts with WEFS microscopy.

Agnarsson et al. [50] picked up WEFS microscopy in 2015 for label-free sensing of gold nanoparticles (AuNPs), vesicles and living cells, and compared it with WEFF results, all on symmetric polymer waveguide structures.

The focus of the following sections will be the WEFF and WEFS studies of the author's group.

1.3   Experimental

1.3.1   Waveguides

Home-made, glass waveguides on fused silica (step-index slab waveguides) or ion exchanged waveguides with holographic coupling gratings will be addressed here [41, 51]. The waveguides were reusable various times after thorough cleaning. Cleaning procedure consisted sonification in 70% ethanol for 20 min followed by blow-dry with nitrogen gas. To remove organic material, the dried samples were cleaned with Nano-Strip (KMG Chemicals Inc., Fremont, CA, USA) at 80 °C for 5 min. After Nano-Strip application, the substrates were extensively rinsed in Milli-Q water and blown dry again [52]. However, with each cleaning cycle the waveguides became thinner and needed characterization before every new experiment.

The samples were designed in a way that the coupling grating is always kept outside the sample area and should not be altered during the experiment to keep the coupling conditions and therefore, the coupling efficiency, constant, Fig. 1.1. This is important because the mode coupling conditions change when material is adsorbed

Fig. 1.1 General WEFF-
and WEFS chip design: the
coupling grating is located
outside the sample area

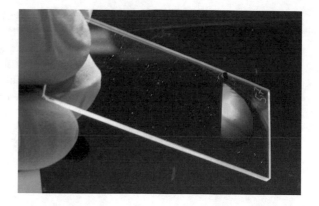

or desorbed at the grating position [53]. In addition, out-coupled photons from the
grating produces artefacts in the WEFF and WEFS images. Any periodic structures
should be avoided in the sample area due to undesired interference effects.

1.3.2 Cell Culture

The cell culture was according with [52] will be transcribed here. Osteoblastic cell
line MC3T3-E1 (subclone 4, ATTC Catalog 3 CRL-2593) culture was carried out
in flasks. Three hours UV light exposure was used to sterilize cleaned waveguides. Growth medium was prepared from 17.8 ml α-minimum essential medium
1X (MEM; Gibco), 2 mL fetal bovine serum (FBS; Gibco) and 0.2 mL antibiotic-
antimycotic solution 100X (Anti-Anti; Gibco). First the medium was aspirated from
the cell culture flask. Dulbecco's phosphate-buffered saline 1X (PBS; Gibco) was
added to wash the cell layer and aspirated subsequently. To detach the osteoblasts
from the vessel, wall, 5 ml trypsin-EDTA (0.05%, Gibco) was added and incubated at
37 °C for 5 min. The cell culture was checked by phase-contrast microscopy in order
to confirm cells release into the suspension. Trypsin neutralization was carried out
by adding 9 mL growth medium. The resulting cell suspension was then diluted in
growth medium to 10,000 cells per mL. The waveguides were placed in a Petri dish
and 1 mL cell suspension was deposited per surface. Samples were then incubated
for 24 h at 37 °C, 100% humidity and 5% CO_2.

After incubation the waveguides were removed from the growth medium and
excess medium was drawn. Each waveguide was then rinsed three times in PBS. For
fixing waveguides with cells on top, were placed in a 4% paraformaldehyde solution
in PBS for 10 min at room temperature. Subsequently, samples were rinsed three
times with PBS. Desiccation was prevented, by keeping the samples in PBS until
further treatment. A 1.5 mg DiO in 1 mL dimethyl sulfoxide (DMSO) solution was
prepared and heated to 37 °C for 5 min. This solution was then sedimented for 5 min
at 2000 rpm in order to separate solid residues. The staining solution was achieved

by taking of 10 µL of this solution and dissolving it in 1 mL of growth medium. A volume of 200 µL this solution was then pipetted onto the corner of each waveguide, which where gently agitated until complete cell coverage by the solution. These samples were left for 20 min in the solution to guarantee the incorporation of the dye. Afterwards, the staining solution was drained and PBS was used to wash the waveguides. For the removal of all unbound dye, the samples were immersed in PBS for 10 min and drained again. The wash cycle was repeated two more times. The waveguides were stored in PBS until performing WEFF microscopy. This procedure delivers fixed cells, cells that are "frozen" in their habitus [52, 54–56] with the dye situated in the plasma membrane of the cells.

1.3.3 Bacterial Culture

The bacterial culture has been previous described [52] will be transcribed here for consistency.

Nitrobacter sp. 263 was cultured on R2A (Difco™) plates at room temperature (approximately 23 °C) for two weeks. For each colonization experiment, bacteria from one R2A plate were removed and suspended in 1 ml of filter-sterilized (0.45 µm pore-size) distilled deionized water to produce an aqueous bacterial suspension (with 10^6 bacteria/ml). A distinct R2B stock solution (i.e., broth/liquid culture medium) was prepared by dissolving R2A in sterile, distilled, deionized water and filtering to remove the agar constituent keeping the dissolved nutrients for bacterial growth.

Attachment of the bacteria to the waveguide surface was attained by placing a 50 µl aliquot of the bacterial suspension on waveguide top for 1 h at 37 °C. After bacteria attachment, the waveguide was rinsed with sterile, distilled water and placed in a sterile Petri dish with 20 ml of R2A and after incubated for 24 h at 37° to allow bacteria to grow. These samples were not agitated. After 24 h incubation, bright field microscopy was used to examine the waveguides in looking for microcolonies formation. Images were taken of live cells in growth medium. WEFS microscopy was then used to analyse the samples. Sterilization experiments of were undertaken. For this separate bacteria suspensions of 10 ml (with 10^6 bacteria/ml) were placed in a sterile, open glass dish and exposed, in a low pressure collimate beam apparatus (LPCB) [52, 57, 58] to doses of 2, 4, 8, 14, 20 and 30 mJ/cm^2 [52, 58]. This sterilization by UV photon was chosen mainly for its ability to disrupt and dimerize neighboring DNA bases (thymine dimerization) that hinders bacterial growth but not viability [52, 59, 60]. The 'sterilized' bacterial suspensions, obtained for the different dose exposures, were used in colonization experiments identical to those described above.

Before the first and second colonization assays, separate 1 mL aliquots of all bacterial suspensions were stained using BacLight™ (Invitrogen) Live-Dead stain and examined using fluorescence microscopy in order to confirm that the cells were viable.

1.3.4 Microscopic Analysis: WEFF and WEFS

An inverted microscope from Zeiss, Oberkochen, Germany, with the waveguide located on the sample stage was used for the WEFF and WEFS microscope assemblies consisted of an inverted microscope (Figs. 1.2 and 1.3). The specimen was placed on the waveguide's top. An argon ion laser (35 LAP 341-200, CVI Melles Griot) working at $\lambda = 488$ nm with a tuneable output power in the 7–126 mW range or a 0.5 mW, 543.8 nm HeNe laser from Research Electro-Optics, were used as light sources in WEFF and WEFS assemblies, respectively. The laser power was reduced by placing a neutral density filter directly behind the laser, avoiding overexposure and bleaching. The beam diameter was controlled by means of an iris aperture. A coupling grating located on the waveguide was used to couple the laser beam in the waveguide to a chosen mode. For the case of WEFF microscopy, undesired excitation wavelength was blocked via a long pass filter with a 490 nm cut-off wavelength (3RD490LP, Omega Optics, Brattleboro, VT), placed between the objective and the camera. The out-coupled intensity at the waveguide end was measured by a large active area photodiode (FDS1010, Thorlabs, Newton, NY) for determining the coupling efficiency when required. A cooled CCD-camera (Pursuit—XS 1.4 Diagnostic Instruments Inc., Sterling Heights, MI, USA), controlled with SPOT 5 Basic (Spot Image Solutions, Sterling Heights, MI, USA) was used for imaging. Bright field microscopy images of the samples were also obtained with the same field of view objective-lens as in the WEFF/WEFS microscopy images and processed through Image Pro Express software facilities (Media Cybernetics, Rockville, MD).

The laser beam coupling can be from below the waveguide chip in parallel to the objective lens as depicted in Fig. 1.2. This has the advantage that all beam steering can be done on the optical table. The disadvantage is the close proximity to the objective lens which restricts the available coupling angle substantially and forbids backward coupling.

In the alternative design the coupling is carried out from the top (Fig. 1.3). The advantage here is a restriction less coupling angle. The disadvantages are the possible

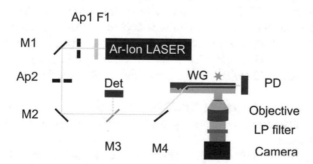

Fig. 1.2 Schematic of WEFF microscope with coupling through the waveguide chip from below: Ap—apertures, F1—neutral density filter, M—mirrors, WG—waveguide and PD—photo diode. For WEFS microscopy a HeNe laser was used and the LP filter omitted [52]

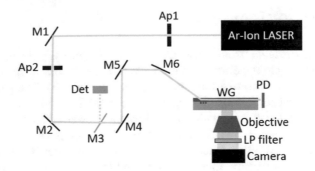

Fig. 1.3 Schematic of a WEFS microscope with coupling into the waveguide from above: Ap—apertures, M—mirrors, WG—waveguide and PD—photo diode. For WEFS microscopy a HeNe laser was used and the LP filter omitted. Optical elements Ap1 to M4 are mounted on the table

reflections and scattered photons from the objective lens, and some of the beam steering needs to be carried out above the optical table making the system more prone to vibration issues.

1.4 TIRF/TIR Versus WEFF/WEFS

Both TIRF/TIR and WEFF/WEFS microscopies assemblies, employ sample illumination trough evanescent fields at the substrate surface which are coming from total internal reflection. In more recent TIRF/TIR microscopes a laser beam is optomechanically guided within the microscope and the objective lens, allowing a laser beam to undergo total internal reflection at a the high refractive index substrate carrying the specimen and is placed above the objective lens. Specially designed objective lenses having high magnification and numerical aperture objectives together with built-in optical path control are required for microscope construction. All angles above critical angle of TIR can be theoretically achieved in this way, allowing the possibility to achieve different penetrations depths by using different angles. This feature can even be used to measure distances from the substrate surface [61]. From the practical point of view, the microscopes are set to particular angles, for achieving both high quality TIRF/TIR imaging and high quality epi-fluorescence or bright field.

Besides, manual operation of a TIRF/TIR microscope can readily give rise to the evanescent mode loss and consequently to a full specimen exposure to the laser beam and sample damage.

Reviewing the literature in particular with respect to the use of TIRF microscopy for distance measurements it can be seen that, besides Burmeister's excellent work by the middle of the 1990s [62], during the development stage of TIRF microscopy, very little has been published on exploring different penetration depths.

In regard to TIR microscopy is performed identically but excluding the dye from the samples and the necessary filter sets. Scattered photons instead of fluorescence photons are collected. Bright field images are taken for comparison since epi-fluorescence is not realizable. It should be pointed out that few distance work involving TIR microscopy have been reported so far [29]. This is not surprising since the scattering intensities are difficult to analyze because all refractive index fluctuations present in the evanescent field contribute to the signal and these are not necessarily controllable, in particular with living cultures producing extracellular matrix in the case of cells or extracellular polymeric substance (EPS) when imaging live bacteria.

In WEFF/WEFS microscopies, the waveguide modes resonances dictate the available evanescent fields and penetration depths. Thus the number of options is limited by the number of modes propagating in the waveguide. In TIRF and TIR microscopies, the evanescent field penetration depth of the is limited to ~200 nm, whereas a waveguide can yield penetrations depths from below 100 nm to over a μm, by refractive index tuning and thickness architecture of both core and cladding layers [44]. Extended illumination area over macroscopic dimensions can also be achieved from planar waveguides which is only limited by the attenuation of the propagating waveguide mode.

In addition, as the beam in WEFF/WEFS cannot escape from the waveguide; the WEFF and WEFS microscopies carry the intrinsic safety mechanism of sample overexposure and damage.

Well characterized waveguides where the evanescent fields and penetration depths are well known can be used for quantitative measurements [43].

WEFF and WEFS microscopies do not seek for state-of-the-art microscopes or objective lenses assemblies. Their technologies can be based on a few simple accessories and attachments to standard inverted microscopes. It is therefore straightforward to image the specimen in any magnification and field of view available due to standard long distance objective lenses by just turning the objective lens revolver without the necessity of beam stirring. Due to the evanescent field formation being taken care of by the substrate and completely independent from the entire microscope, different field of views or magnifications still deal with the same illumination conditions allowing direct comparison of images or measurements after changing magnification and field of view.

By enhancing the image acquisition integration time, epi-fluorescence images can be obtained. This is due to waveguide slightly scattering, which provides 3D excitation or scattering photons to the specimen.

Comparing TIRF and WEFF images of the same samples identical image information can be observed [40]. Both are diffraction limited, in such a way that lateral resolution is dependent on the working laser wavelength and highest magnification lens and its numerical aperture (NA) supported by the microscope used. Resolution in the z-direction (normal to the substrate) is about 7 nm for both types of microscopes.

For WEFF and WEFS microscopies wide use it is necessary to have easy access to and supply of inexpensive waveguide substrates. Thus, it is necessary to implement a mass producible waveguide-chip.

1.5   Multimode Waveguide Use to Static Distance Mapping

A 651 ± 2 nm thick waveguide with refractive index of $n = 1.840 \pm 0.001$ was used for distances mapping of a dye, located in the plasma membrane of fixed osteoblasts. For simulating the evanescent fields the volume above the waveguide was assumed to be water with a refractive index of 1.33. Two images taken with the TM_1 and TM_2 mode were used to calculate the dye distance map [52, 63].

The WEFF image in Fig. 1.4 depicts four osteoblasts well spread and indicating the nuclei and some cell extensions. The dye distance map shows lower distance-colors (blue to yellow) in the cells areas from in the range of 0, to ~130 nm. In the unoccupied area, the unstained medium, where the raw data do not show fluorescence, only noise is present. This is depicted as distances in the order of the penetration depth of the evanescent field i.e. $\sim 160 \pm 40$ nm (red pixels with yellow). In addition, isolated spots in the no-sample area (outside the cells) are seen in very dark blue. These spots are correlated to un-physical distance values below zero caused by microscopic damages of the waveguide. These un-physical distances should always be omitted in image interpretation. All four osteoblasts can be found in the distance map and cell outlines are similar to the cells depicted in the "epi-fluorescence" image. It should be noted that filopodia and the thinly spread cell body are more clearly seen in the distance map. The distance map (Fig. 1.4) does not depict any information about the nuclei. Not the entire cell body reached down very close to the surface, as expected. At some of the cells' outer lines and at some extreme tips of the spread cells, small regions of only a few pixels in diameter were found with distances of ~10–25 nm, typical of a focal adhesion [4, 64]. Twice line like accumulations of dense focal adhesions are found (blue lines with distances around 10–25 nm). Between the focal adhesions, there are regions in lighter blue depicting distances around 40–50 nm as well as greenish areas depicting distances around 70–80 nm. Filopodia of the cells, which are very faintly seen in the epi-fluorescence images, are clearly visible in the distance map as thin spikes with a blue (possible focal adhesions or point contacts) or green (possible extracellular matrix contacts) center and green-yellow surroundings [64].

Figure 1.5 depicts one well spread osteoblast taken in epi-fluorescence WEFF mode and false color distance map [49]. Two z-cuts through the distance map have been made: one randomly through the cell, (Fig. 1.5c) and one through an area including the smallest distances of the cell (Fig. 1.5d) [52]. The area outside the cell revealed to be nearly homogeneously dark red colored. The noise level in the no-sample regions is clearly seen in the z-cut data; it is the noisy data at an average distance of ~90 nm on both cell sides [52]. The cell itself is shown by the depressions in the z-cuts with the dips indicating adhesions. The spreading of the cell is excellently depicted by the distance map.

The cell is attached at all extreme spreading points, however not necessarily as focal adhesions since, distances above 40 nm and up to 50 nm, possible close contacts, are found. In the cell center, focal adhesions can be seen.

From the z-cuts the position of the plasma membrane/dye location along the cut line in nm can be seen. For the random 'c' cut, three "small" distances in the order

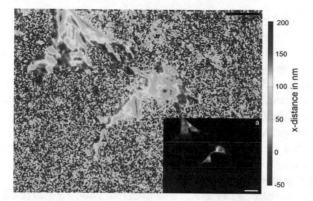

Fig. 1.4 Dye distance map with four osteoblasts, false color representation. The inset represents a WEFF image with increased integration time of the same field of view. Both scale bars represent 50 μm [52]

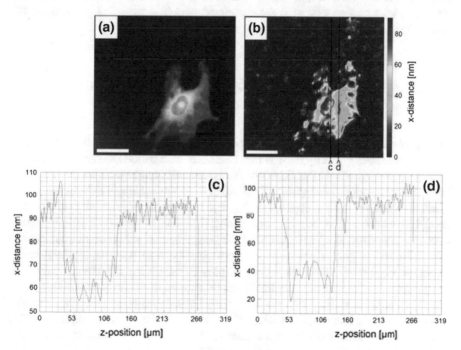

Fig. 1.5 Single osteoblast imaging results: **a** epi-fluorescence WEFF image, **b** dye distance map false color representation, **c** z-cut through cell at random position 'c' in part (**b**) and **d** z-cut through cell at smallest distance locations at position 'd' in (**b**). The cuts in (**b**) from bottom to top are represented in (**c**) and (**d**) from left to right. The scale bars represent 25 μm [52]

of ~55 nm are found, as well as a couple of more bends towards the substratum with distances of ~62–67 nm [52]. The maximum heights of the plasma membrane from the waveguide surfaces between the bends towards the substratum are found to be between 62 and 75 nm.

In the z-cut 'd' through the small distance adhesions one 18 nm focal adhesion can be found as well as 25–35 nm distance contacts. The maximum heights of the plasma membrane from the waveguide surfaces in this case are 37 and 45 nm. The bending of the membrane towards the cytoplasm between these adhesions points is clearly depicted. The relative straight lines between the "maxima and minima" in the distance curve bear a resemblance to a stretched rubber band. One needs to keep in mind that the surface tension of the plasma membrane tries to minimize the surface area, trying to force the cell into a spherical shape. The adhesions are obvious biological disruptions of the physical effect of surface minimization.

It would be interesting to monitor and quantitatively analyze the dynamics of a living cell moving and forming lamellipodia and new adhesions as well as retrieving lamellipodia and withdraw adhesion. With the current set-up, time laps distance mapping it is not yet possible. An automated, motorized mirror adjustment for M4 (Fig. 1.2) with a feedback loop for optimized coupling from the photodiode PD (Fig. 1.2) needs to be implemented.

1.6 Cell Plasma Membranes Dynamic Solubilisation Studies

Detergent-membrane interactions have been the subject of many studies [65]. Functional membranes typically exist in the fluid state, also called the liquid-disordered state. Due to difficulties of working with authentic cell membranes, simplified membrane models—such as supported lipid bilayers or liposome mimicking biological systems—have often been used to investigate detergent-membrane interactions [65]. Model membranes were helpful in exploring the basic membrane functions. However, in comparison to a living cell, with integral and peripheral proteins, cholesterol molecules and oligosaccharides in and on their plasma membrane, artificial membrane models cannot mimic all aspects of plasma membrane function. In addition, studying the interaction between lipids and detergents in the form of vesicles (liposomes) or supported lipid bilayers has several other disadvantages. For example, in supported lipid bilayers, the quality of the deposited film plays a major role. The direct contact with the underlying substrate affects the bilayer's structure and fluidity, and blocks access of solutions to both sides of the membrane.

The results of lipid-detergent interaction studies using bio-membrane models have been related to a three-stage model, which was described by Lichtenberg et al. [66]. In stage I, with increasing detergent concentration, detergent incorporates into the bilayer. At this stage, solubilization does not occur, but the bilayer becomes saturated with detergent. At stage II, with further increase in detergent concentration, the bilayer starts to solubilise. Lipid vesicles saturated with detergent form and

Fig. 1.6 Three cells normalized integrated intensities versus time. Triton X-100 (0.013 w/w%) addition is indicated by the arrow [52]

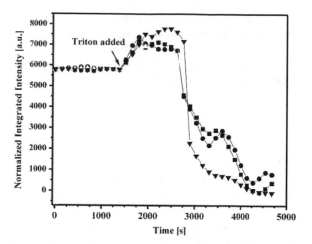

coexist with mixed micelles of lipid and detergent. At stage III, the entire membrane solubilises, and only mixed micelles exist [67, 68].

Osteoblast were cultured on the waveguides and imaged alive with time laps WEFF microscopy. At a certain time Triton X-100 was added to the medium to start solubilisation. Figure 1.6 shows the normalized integrated intensity of the WEFF fluorescence signal of three example cells imaged with time.

Without detergent, the integrated intensities are constant indicating negligible photo bleaching. In the presence of detergent, three reproducible kinetic stages can be seen: (i) an increase in fluorescence intensity, (ii) a plateau, and (iii) a decrease in intensity. Therefore, a comparison to or an adaption of the established three-stage model is possible. In stage I, the membrane takes up detergent and the concentration of detergent rises in the plasma membrane. The integrated fluorescence intensity increases due to suppression of fluorophore quenching by dilution of the dye with detergent [69] in the cell membrane. In this stage, solubilisation does not occur. According to the model, stage I ends when the membrane becomes saturated with detergent [42]. The end of stage I is seen in Fig. 1.6 when the intensity increase ends and the plateau starts.

In stage II, where artificial membrane solubilisation takes place, the detergent-saturated lipid bilayer undergoes a structural transition and converts partially into lipid-detergent mixed micelles; however, these micelles are not yet mobile, but still incorporated in the membrane. So, stage II can be seen as the plateau in which intensity remains constant as the dye is not leaving the evanescent field. At this time, the dye is still located either in the membrane or in formed micelles in unquenched conditions mixed with detergent.

During stage III, the micelles become mobile and leave the evanescent field, leading to a decrease in integrated intensity. Individual micelles are too small to be seen with the WEFF microscope.

By changing the Triton X-100 concentration the duration of all three phases changed: the higher the detergent concentration the quicker the solubilisation stages [42].

WEFF microscopy confirmed that living osteoblasts are solubilized in the same way as model membranes.

1.7 Time Response of Osteoblasts to Trypsin

Trypsin is a serine protease and cleaves peptide chains. Therefore, trypsin is used in laboratories to cleave proteins bonding the cultured cells to the dish, so that the cells can be suspended in fresh solution and transferred to fresh dishes.

Healthy osteoblast cells were grown directly on the waveguide and monitored with time laps WEFF microscopy. Trypsin was used at 0.05 and 0.02% concentration. Upon addition of 0.05% trypsin, the cells were lifted very fast and only individual focal adhesions could be imaged. However, with the lower concentration changes in cell morphology could be observed, such as cell retraction.

The quick disappearance of an individual adhesion point at the high trypsin concentration was examined. The focal adhesion point had the appearance of a bright circular dot. A series of images were taken with time and analyzed. Figure 1.7 depicts the kinetic behaviour of the adhesion point's disappearance, with respect of its integral intensity and size. Clearly both the size and the integral intensity of this individual focal adhesion point decreased in an S-shaped curve and provided basically identical kinetic information about the detachment of the cell.

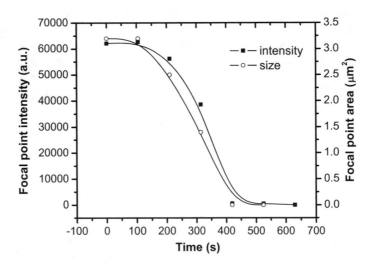

Fig. 1.7 The impact of a 0.05% trypsin containing medium on an individual focal adhesion: intensity and size decrease with time. The lines are guides to the eye [52]

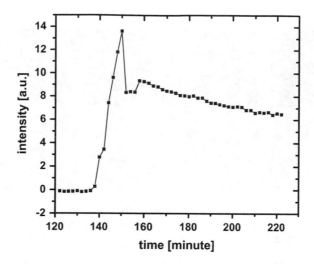

Fig. 1.8 Integrated, intensity of 5 individual re-appearing adhesion points after exchanging a trypsin-containing medium at $t = 0$ to a trypsin-free medium [52]

A sample was treated with 0.02% trypsin. The cells have shown cell retraction, and detached from the surface, leaving a black feature less evanescent image. After the trypsin treatment the medium was exchanged carefully to a trypsin-free environment. The imaging was continued. The osteoblasts, still alive, re synthesise new adhesion proteins for the formation of new adhesion points. The kinetics of the adhesion process, unit the cell population died at around 150 min and lost adhesion again, is depicted in Fig. 1.8.

1.8 Bacteria Sterilization

Studies on the attachment of bacteria onto surfaces using WEFS microscopy detection is a quick method for investigations regarding bacterial sterilization treatment [49]. We hypothesized that non-potent, sterilized cells do not attach to surfaces and do not form microcolonies. Therefore, we have treated identical bacteria sample batches with different UV doses (2, 4, 8, 14, 20 and 30 mJ/cm^2). After the UV illumination the viability was measured. The UV illumination did not result in bacterial death. As a control, one sample was left without UV treatment. All bacteria illuminated with different UV doses and the control were cultured identically and examined using WEFS microscopy after 24 h. Figure 1.9 shows a series of WEFS and bright field images of the control and UV treated bacteria.

The relative signal attributed to attached colonies and individual bacteria on the waveguide surface decreased as exposure to UV illumination was increased (Fig. 1.9). It is significant to note that the highest dose of 30 mJ/cm^2 was not sufficient to completely prevent bacterial attachment. Both WEFS and bright field microscopy

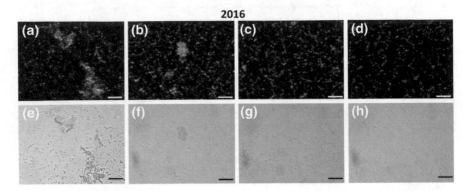

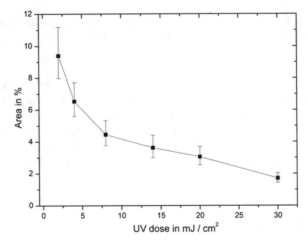

Fig. 1.9 WEFS and bright field microscopy images of UV illuminated, sterilized bacteria after 24 h of culturing: **a** and **e** control: 0 mJ/cm^2, **b** and **f** 8 mJ/cm^2, **c** and **g** 20 mJ/cm^2, **d** and **h** 30 mJ/cm^2. The scale bar is 50 μm [52]

Fig. 1.10 Percentage of occupied area of bacteria versus applied UV dose. The line is a guide to the eye only

demonstrated that the highest dose resulted in the attachment of primarily individual bacteria, demonstrating that while attachment still occurred with increasing UV-dose, microcolony formation was prevented.

In order to yield quantitative data, a Matlab program was written to investigate the intensity distribution of each WEFS image and to calculate the percentage of area (i.e., pixels with signals above the defined threshold) occupied by bacteria (i.e., individual cells and cells comprising distinct colonies). Figure 1.10 shows the percentage of area on a sample occupied by bacteria versus the applied UV dose. Although the percentage of surface area with attached bacteria was decreasing exponentially, it did not reach zero. Bacteria were still attached to the waveguide surface despite the UV treatment. A "safe"-dose can be extrapolated by the data.

1.9 Cell Granularity and Adhesions

Fixed osteoblasts were imaged with WEFS microscopy. Figure 1.11 shows a bright field image of a single osteoblast and the corresponding WEFS image. In the WEFS image the nucleus can be located: it is the dark area in the cell center. In addition, the granular structures in the cell body and the adhesion sites at the cell outline were visible. Figure 1.11 indicates with the arrow the propagation direction of the waveguide mode. The cell's boundary first hit by the propagating light was shown very clear and with many adhesions points. The other three outer lines depict the adhesion points but not the complete cell boundary. At this chosen integration time the WEFS image depicts adhesions due to the evanescent illumination and the cell granularity due to 3D scattering of the waveguide.

Cell-substrate adhesions could be distinguished from scattering centers located further away from the substrate, the granularity of the cell, by varying the integration time. This is shown in Fig. 1.12.

With a very short integration time only a few spots appeared on the image in the areas where the cell was well spread. These spots are the cell's adhesions within the evanescent field. With increasing integration time, more and more features appeared, such as the cell nucleus area, the cells boundary and the cell granularity.

These experiments showed that not necessarily fluorescence staining was needed for imaging focal adhesions and hence getting some cell-substratum interaction measures. As in WEFF microscopy larger integration times lead to 3D information of the cell. Further detailed analysis, e.g. whether WEFS data are comparable with flow cytometry (scattering mode), need to be done.

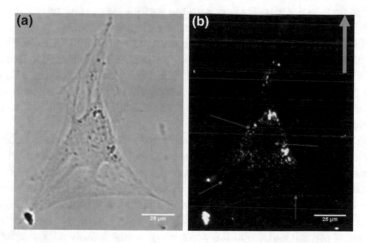

Fig. 1.11 a Bright field image and **b** WEFS image of an osteoblast taken with an exposure time 3000 ms. The green arrow indicates the direction of light propagation. The scale bars are 25 μm [52]

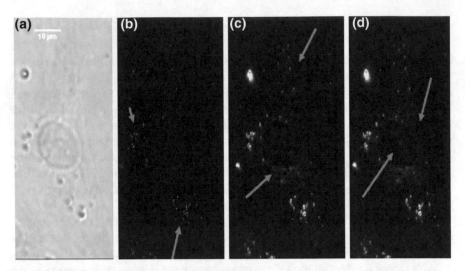

Fig. 1.12 **a** Bright field image of a single osteoblast and **b–d** corresponding WEFS images with integration times of 500 ms, 1000 ms and 1500 ms, respectively. The arrows point to the features mentioned in the text: (**b**) adhesions, (**c**) granularity and cell boundary, and (**d**) nucleus and cell boundary [52]

1.10   Imaging with WEFF/WEFS Microscopy

For evanescent imaging with WEFF and WEFS microscopy on the same sample, osteoblast cells were fixed and stained. The aim was to determine the information differences in the images taken with the various evanescent microscopy forms at identical samples and integration times. It is possible to image a stained object with WEFF microscopy implementing the long pass filter to block excitation light, with WFFS/WEFS combination microscopy by collecting both scattered excitation light and fluorescence photons, and with WEFS microscopy applying a short pass filter blocking the fluorescent emission wavelengths. Figure 1.13 shows a series of images of a single osteoblast taken with bright field, WEFF, WEFF/WEFS and WEFS microscopy.

Figure 1.13a shows the bright field microscopy image of a single osteoblast cell. The nucleus, and the outline of the cell were clearly visible in this image. To confirm the visualization of the entire cell with both WEFF and WEFS methods, an integration time of 6000 ms was used. The WEFF microscopy image is shown in Fig. 1.13b. Both the cell outline and cell body was distinctly visible in the WEFF image. The nucleus appeared black with some structure in it. The cell body was distinguishable from the other parts of the cell because of the presence of many densely packed bright spots around the nucleus. The outline of the cell which is actually the spread region of the cell was less bright than the rest of the cell but still unmistakable as it was identical to the bright field image of Fig. 1.13a. The white arrow in this image refers to the regions where the cell is touching an adjacent cell. The waveguide implemented here

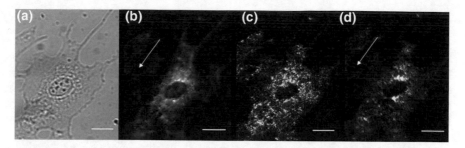

Fig. 1.13 Osteoblast imaged with a TM mode and 6000 ms integration time: **a** bright field image, **b** image captured with WEFF microscopy with a 560 nm long pass filter blocking excitation light of 543 nm, **c** WEFF/WEFS image captured without filters; hence both scattered and emission photons of the dye form the image, and **d** image captured with WEFS microscopy with a 550 nm short pass filter blocking the fluorescence. The arrows indicate where the cell is touching an adjacent cell. Scale bars represent 20 μm

has an evanescent field of 100 nm. Only the close contact regions of the cell should be visible in a "real" WEFF image. But due to the high integration time, the entire cell became visible as an epi-fluorescence image where parts of the cells located far away from the surface could also be seen. Close contact regions and focal adhesions are not possible to visualize with this types of WEFF imaging strategy.

Figure 1.13c shows the cell captured with no filters in WEFFS/WEFS combination mode. As a result, the scattered photons and the photons from the dye emission are forming the image. Although the cell outline was visible and nucleus distinguishable, there was too much intensity present in the entire cell making it impossible to distinguish the spread region of the cell and the cell body. Also the granularity was not visible as expected (Fig. 1.11).

Figure 1.13d, was captured with pure WEFS microscopy employing a 550 nm short pass filter blocking the fluorescence. This image showed the outline of the cells and the nucleus, but no clear distinguishing between the spread region of the cell and the cell body was possible. The links to the touching neighbor cell indicated by an arrow, were less prominent compared to the epi-fluorescence WEFF microscopy image. The granularity was distinctively different from the images of the cells which were not stained (Fig. 1.11).

From these studies it became clear that both methods have their individual optimum microscopy settings for achieving informative images. The amount of scattered light increased when a cell is stained and does not necessary deliver the same information about granularity as an unstained cell. Unresolved dye might act as additional scattering centers delivering artefacts. For an informative comparison of WEFF and WEFS images two sample sets should be prepared and imaged with individual optimized conditions (microscope settings and sample preparation).

1.11   All-Polymer-Waveguide-Chips

In order to allow WEFF and WEFS microscopy to be used by the interested communities, typically biophysics, biology, biochemistry and medical laboratories, but also for coating engineering (aging studies, homogeneity studies, anti-microbial tests, etc.) the waveguide chips need to be (commercially) available and at a reasonable cost. Mass production is the only way to accomplish this. An all-polymer-waveguide-chip with an imprinted coupling grating is one way to achieve this goal.

The all-polymer-waveguide chip was designed on the basis of a PMMA substrate.

The imprinting was performed into the PMMA with a home-fabricated silicon stamp, and in a subsequent step a polystyrene waveguide was spin coated on top.

Figure 1.14 shows an SEM image of an imprinted grating with a periodicity of 670 nm and a depth of 200 nm.

First experiments with the all-polymer-waveguide-chips have produced promising WEFF imaging results (Fig. 1.15). The WEFF image of the HeLa cell on the all-polymer-waveguide-chip does still look like an epi-fluorescence image. Also the distinct spotty or dotty pattern due to the adhesions at the end of the filopodia and at the outer rim of the cell are still missing. The polymer chips still suffer from too many scattered photons due to too many imperfections in the waveguide. The waveguide spin coating conditions need to be improved (dust free and with a homogeneous thickness throughout the sample). In addition, development towards mass produced chips is necessary. The grating of Fig. 1.11 and the all-polymer-waveguide-chip of Fig. 1.14 were fabricated by imprinting one grating into one PMMA substrate. The imprinting and waveguide spinning procedure needs to be scaled up to fabricate 16, 25, 36 or more chips in parallel on one substrate with one imprinting procedure and a subsequent high quality spin coating process and subsequent separation of the individual chips.

Fig. 1.14 SEM image of a PMMA imprinted coupling grating. The periodicity is 670 nm [52]

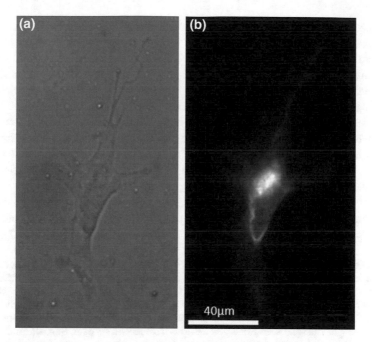

Fig. 1.15 All-polymer-waveguide-chip: **a** Bright field image and **b** WEFF image of a HeLa cell stained with DiO [52]

1.12   Conclusions and Outlook

A simple method to perform TIR and TIRF microscopy with a conventional inverted optical microscope by implementing an optical slab waveguide as the illumination source was discussed as well as the suite of advantages a confined beam in a waveguide offers in comparison to a standard TIR(F) microscope.

Both WEFF and WEFS microscopies were applied to a variety of biophysical questions: simple imaging of adhesions, quantitative investigations such as dye distance mapping and analyzing kinetic phenomena. A critical analysis of images taken in a WEFF/WEFS combination found that the combination is not recommendable. Each method should be used on samples especially prepared and with the optimum individual microscopy and image acquisition conditions.

In order to make the technology available for an interested scientific community, the availability of the waveguide-chip is essential. Therefore, a methodology for the fabrication of a mass produced, cost-effective waveguide-chip based on polymers only was developed and tested. In the future, the all-glass-chips should come with a surface functionalization allowing reusability. Time laps distance mapping is not possible yet, but planned.

Various types of interface and surface related biophysical and biological questions can be addressed with WEFF/S microscopy. They carry in addition the opportunity of an implementation in senor technology. Various options have been published already [45, 50].

There is a huge opportunity to also use WEFF/S for advanced measurements. The WEFF microscope can be simultaneously operated by propagating modes at different wavelength or directions for any kind of pump probe or resonance experiment, or a sensing scheme based on a Förster transfer in a dye upon binding of an analyte. Pulsed laser operation is another option. The scattering microscopy is responsive to any changes in the size or the refractive index (density) of the scattering entity within the evanescent field. The monitoring sensitivity of surface recognition reactions could easily be enhanced by increasing the scattering power by a gold nanoparticle [70] or by increasing the size of a scattering entity due to the binding [50].

Silane chemistry will allow to tune the waveguide's surface functionalization, both for all-glass and all-polymer-chips. Hydroxyl groups can easily be produced by oxygen plasma or UV ozone treatment for further functionalization [71, 72].

Acknowledgements Many co-workers, students, PDFs and colleagues are thanked for their contribution in the past, the present and the future for developing and applying WEFF and WEFS microscopy: Frank Thoma, Uwe Langbein, John J. Armitage, Huge Trembley, Michael Nietsche, Abdollah Hassanzadeh, Elisabeth Pruski, Jeffrey S. Dixon, Stephen Sims, Rebbeca Stuchburry, Sabiha Hacibekiroglu, Daniel Imruck, Christopher Halfpap, Michael Morawitz, Qamrun Nahar, Darryl K. Knight, Susanne Armstrong, Jeremia Shuster, Frederik Fleisser, Mihaela Stefan, Kibret Mequanint, Beth Gillies, Rene Harrison, Gordon Southam, Cheryle Seguin, Hong Hong Chen, Donglin Bai, Douglas Hamilton and Rony Sharon.

References

1. X.F. Niu, Y.L. Wang, Y.L. Luo, J. Xin, Y.G. Li, J. Mater. Sci. Technol. **21**, 571–576 (2005)
2. H. Storrie, M.O. Guler, S.N. Abu-Amara, T. Volberg, M. Rao, B. Geiger, S.I. Stupp, Biomaterials **28**, 4608–4618 (2007)
3. J.S. Burmeister, L.A. Olivier, W.M. Reichert, G.A. Truskey, Biomaterials **19**, 307–325 (1998)
4. W.T. Chen, S.J. Singer, J. Cell Biol. **95**, 205–222 (1982)
5. J.S. Burmeister, G.A. Truskey, J.L. Yarbough, W.M. Reichert, Biotechnol. Prog. **10**, 26–31 (1994)
6. K.F. Giebel, C. Bechinger, S. Herminghaus, M. Riedel, P. Leiderer, U. Weiland, M. Bastmeyer, Biophys. J. **76**, 509–516 (1999)
7. H. Verschueren, J. Cell Sci. **75**, 279–301 (1984)
8. B. Braun, P. Fromherz, Appl. Phys. A-Mater. Sci. Process. **65**, 341–348 (1997)
9. E. Atilgan, B. Ovryn, Curr. Pharm. Biotechnol. **10**, 508–514 (2009)
10. M. Abercrombie, J.E. Heaysman, S.M. Pegrum, Exp. Cell Res. **67**, 359–367 (1971)
11. K.E. Sapsford, L.C. Shiver-Lake, in *Principles of Bacterial Detection: Biosensors, Recognition Receptors and Microsystems*, ed. by M. Zourob, S. Elwary, A.P.F. Turner (Springer, New York, 2008), pp. 109–123
12. J. Pizarro-Cerda, P. Cossart, Cell **124**, 715–725 (2006)
13. J.D. Oliver, J. Microbiol. **43**, 93–100 (2005)
14. A.E. Madkour, G.N. Tew, Polym. Int. **57**, 6 (2008)
15. A.E. Madkour, J.M. Dabkowski, K. Nusslein, G.N. Tew, Langmuir **25**, 1060–1067 (2009)
16. W.E. Stamm, Ann. Int. Med. **89**, 764–769 (1978)
17. E.M. Hetrick, M.H. Schoenfisch, Chem. Soc. Rev. **35**, 780–789 (2006)

18. G.E. Christianson, 2004, in *Molecular Adhesion and Its Applications, The Sticky Universe*, ed. by K. Kendal, pp. 275–303
19. K. Vasilev, J. Cook, H.J. Griesser, Expert Rev. Med. Devices **6**, 553–567 (2009)
20. L. Pires, K. Sachsenheimer, T. Kleintschek, A. Waldbaur, T. Schwartz, B.E. Rapp, Biosens. Bioelectron. **47**, 157–163 (2013)
21. N.P. Pera, A. Kouki, S. Haataja, H.M. Branderhorst, R.M.J. Liskamp, G.M. Visser, J. Finne, R.J. Pieters, Org. Biomol. Chem. **8**, 2425–2429 (2010)
22. A.D. Taylor, J. Ladd, J. Homolas, S. Jang, in *Principles of Bacterial Detection: Biosensors, Recognition Receptors and Microsystems*, XXXII, ed. by M. Zourob, S. Elwary, A. Turner (Springer, 2008), pp. 83–108
23. M. Schmidt, M.K. Hourfar, S.-B. Nicol, A. Wahl, J. Heck, C. Weis, T. Tonn, H.-P. Spengler, T. Montag, E. Seifried, W.K. Roth, Transfusion **46**, 1367–1373 (2006)
24. I.R. Cooper, S.T. Meikle, G. Standen, G.W. Hanlon, M. Santin, J. Microbiol. Methods **78**, 40–44 (2009)
25. M. Zourob, S. Mohr, B.J.T. Brown, P.F. Fielden, M.B. McDonnell, N.J. Goddard, Biosens. Bioelectron. **21**, 293–302 (2005)
26. A. Mazhorova, A. Markov, A. Ng, R. Chinnappan, O. Skorobogata, M. Zourob, M. Skorobogatiy, Opt. Express **20**, 5344–5355 (2012)
27. A. Bauereiss, O. Welzel, J. Jung, S. Grosse-Holz, N. Lelental, P. Lewczuk, E.M. Wenzel, J. Kornhuber, T.W. Groemer, Traffic **16**, 655–675 (2015)
28. X. Liu, E.S. Welf, J.M. Haugh, J. R. Soc. Interface **12**, 20141412 (1–11) (2015)
29. L.V. Smith, L.K. Tamm, R.M. Ford, Langmuir **18**, 5247–5255 (2002)
30. M.A.S. Vigeant, M. Wagner, L.K. Tamm, R.M. Ford, Langmuir **17**, 2235–2242 (2001)
31. G.D. Byrne, M.C. Pitter, J. Zhang, F.H. Falcone, S. Stolnik, M.G. Somekh, J. Microsc. **231**, 168–179 (2008)
32. F. Thoma, U. Langbein, S. Mittler-Neher, Opt. Commun. **134**, 16–20 (1997)
33. F. Thoma, J.J. Armitage, H. Trembley, B. Menges, U. Langbein, S. Mittler-Neher, Proc. SPIE **3414**, 242–249 (1998)
34. T.S. Hug, J.E. Prenosil, M. Morbidelli, Biosens. Bioelectron. **16**, 865–874 (2001)
35. T.S. Hug, J.E. Prenosil, P. Maier, M. Morbidelli, Biotechnol. Bioeng. **80**, 213–221 (2002)
36. T.S. Hug, J.E. Prenosil, P. Maier, M. Morbidelli, Biotechnol. Prog. **18**, 1408–1413 (2002)
37. R. Horvath, H.C. Pedersen, N. Skivesen, D. Selmeczi, N.B. Larsen, Appl. Phys. Lett. **86**, 071101 (2005)
38. H.M. Grandin, B. Städler, M. Textor, J. Vörös, Biosensens. Bioelectron. **21**, 1476–1482 (2006)
39. A. Hassanzadeh, M. Nitsche, S. Mittler, S. Armstrong, J. Dixon, U. Langbein, Appl. Phys. Lett. **92**, 233503 (2008)
40. A. Hassanzadeh, M. Nitsche, S. Armstrong, N. Nabavi, R. Harrison, S.J. Dixon, U. Langbein, S. Mittler, Biomed. Opt. **15** 036018-1–036018-7 (2010)
41. A. Hassanzadeh, S. Mittler, Opt. Eng. **50**, 071103 (2011)
42. A. Hassanzadeh, H. Kan Ma, S.J. Dixon, S. Mittler, Biomed. Opt. **17**(076025), 1–7 (2012)
43. A. Hassanzadeh, S. Armstrong, S.J. Dixon, S. Mittler, Appl. Phys. Lett. **94**, 033503 (2009)
44. B. Agnarsson, S. Ingthorsson, T. Gudjonsson, K. Leosson, Opt. Express **17**, 5075–5082 (2009)
45. B. Agnarsson, J. Halldorsson, N. Arnfinnsdottir, S. Ingthorsson, T. Gudjonsson, K. Leosson, Microelectron. Eng. **87**, 56–61 (2010)
46. H. Keshmiri, B. Agnarsson, K. Leósson, *SPIE*, vol. 8090, 80900 D1-6 (2011)
47. B. Agnarsson, A.B. Jonsdottir, N.B. Arnfinnsdottir, K. Leosson, Opt. Express **19**, 22929–22935 (2011)
48. K. Leosson, B. Agnarsson, Micromachines **3**, 114–125 (2012)
49. Q. Nahar, F. Fleissner, J. Shuster, M. Morawitz, C. Halfpap, M. Stefan, G. Southam, U. Langbein, S. Mittler, J. Biophotonics **7**, 542–551 (2014)
50. B. Agnarsson, A. Lundgren, A. Gunnarsson, M. Rabe, A. Kunze, M. Mapar, L. Simonsson, M. Bally, V.P. Zhdanov, F. Höök, ACS Nano **9**, 11849–11862 (2015)
51. C. Halfpap, M. Morawitz, A. Peter, N. Detrez, S. Mittler, U. Langbein, *DGaO Proceedings*, 0287-2012 (2012)

52. S. Mittler, *Proceedings of the 4th International Conference on Photonics, Optics and Laser Technology, PHOTOPTICS 2016* (SciTePress, 2016), pp. 201–212
53. D.C. Adler, T.H. Ko, J.G. Fujimoto, Speckle reduction in optical coherence (2004)
54. R. Horvath, J. Vörös, R. Graf, G. Fricsovszky, M. Textor, L.R. Lindvold, N.D. Spencer, E. Papp, Appl. Phys. B **72**, 441–447 (2001)
55. L.L. Lanier, N.L. Warner, J. Immunol. Methods **47**, 25–30 (1981)
56. J.W. Smit, C.J.L.M. Meijer, F. Decay, T.M. Feltkamp, J. Immunol. Methods **6**, 93–98 (1974)
57. J.-W. Su, W.-C. Hsu, J.-W. Tjiu, C.-P. Chiang, C.-W. Huang, K.-B. Sunga, J. Biomed. Opt. **19**, 075007 (2014)
58. R.P. Hedrick, B. Petri, T.S. McDowell, K. Mukkatira, L.J. Sealey, Dis. Aquat. Org. **74**, 113–118 (2007)
59. J. Kuo, M. Asce, C.-L. Chen, M. Nellor, J. Environ. Eng. **129**, 774–779 (2003)
60. M. Berney, F. Hammes, F. Bosshard, H.-U. Weilenmann, T. Egli, Appl. Environ. Microbiol. **73**, 3283–3290 (2007)
61. B. Durbeej, L.A. Eriksson, J. Photochem. Photobiol. A: Chem. **152**, 95–101 (2002)
62. G.A. Truskey, J.S. Burmeister, E. Grapa, W.M. Reichert, J. Cell Sci. **103**, 491–499 (1992)
63. J.S. Burmeister, G.A. Truskey, W.M. Reichert, J. Microsc.-Oxford **173**, 39–51 (1994)
64. F. Fleissner, M. Morawitz, S.J. Dixon, U. Langbein, S. Mittler, J. Biophotonics 1–12 (2014)
65. N. Tawil, E. Wilson, S. Carbonetto, J. Cell Biol. **120**, 261–271 (1993)
66. V.N. Ngassam, M.C. Howland, A. Sapuri-Butti, N. Rosidi, A.N. Parikh, Soft Matter **8**, 3734–3738 (2012)
67. D. Lichtenberg, J. Robson, E.A. Dennis, Biochem. Biophys. Acta **821**, 470–478 (1985)
68. G. Csucs, J.J. Ramsden, Biochem. Biophys. Acta **1369**, 304–308 (1998)
69. A. Helenius, K. Simons, Biochem. Biophys. Acta **415**, 29–79 (1975)
70. J.R. Silvius, Annu. Rev. Biophys. Biomol. Struct. **21**, 323–348 (1992)
71. A. Klein, Diploma Thesis, RheinMain University, Rüsseleheim, Germany (2008)
72. S. Kandeepan, J.A. Paquette, J.B. Gilroy, S. Mittler, Chem. Vap. Depos. **21**(275), 280 (2015)

Chapter 2
Characterization of Micro-lenslet Array Using Digital Holographic Interferometric Microscope

Varun Kumar and Chandra Shakher

Abstract When laser light is transmitted through a transparent micro-lenslet array, a phase shift is induced in the transmitted wavefront, depending on the height variation and refractive index of the micro-lenslet array. In this paper, digital holographic interferometric microscope (DHIM) with Fresnel reconstruction method is demonstrated for the characterization of micro-lenslet array. Measurement of diameter (D), sag height (h), radius of curvature (ROC), focal length (f) and shape of micro-lenses are presented in the paper. The height profile of micro-lenses measured by DHIM is compared with commercially available Coherence Correlation Interferometer (CCI) from Taylor Hobson Ltd. UK with axial resolution 0.1 Å. The root mean square error (RSME) between the measurement carried out by DHIM and CCI is 0.12%. The advantage of using the DHIM is that the distortions in the wavefronts due to aberrations in the optical system can be avoided by the interferometric comparison of reconstructed phase with and without the micro-lenslet array.

2.1 Introduction

Micro-optical components such as micro-lenses and micro-lenslet array have numerous engineering and industrial applications for example collimation of laser diodes, imaging for sensor system (CCD/CMOS, document copier machines etc.), for making homogeneous beam for high power lasers, a critical component in Shack-Hartmann sensor, fiber coupling and optical switching in communication systems [1–4]. Also micro-optical components have become alternative to bulk optics for applications where miniaturization, reduction of alignment and packaging cost are detrimental [5]. Therefore, there is a high demand to meet the best

V. Kumar · C. Shakher (✉)
Laser Applications and Holography Laboratory, Instrument Design Development Centre,
Indian Institute of Technology Delhi, Hauz Khas, New Delhi 110016, India
e-mail: cshakher@iddc.iitd.ac.in

V. Kumar
e-mail: varunphy@gmail.com

© Springer Nature Switzerland AG 2018
P. A. Ribeiro and M. Raposo (eds.), *Optics, Photonics and Laser Technology*, Springer
Series in Optical Sciences 218, https://doi.org/10.1007/978-3-319-98548-0_2

quality for micro-optics manufacturing so that needed quality of required system can be achieved. For quality assurance of micro-optics as micro-lenslet array, an economical, accurate, fast, simple and robust measurement technique is needed. Different methods have been developed for micro-optics testing such as mechanical stylus profilometry [6–9], various electron microscopy techniques [9–12], confocal microscopy [9, 13, 14], interferometric methods [9, 15–23], optical coherence tomography (OCT) [24, 25] and digital holographic microscopy [23, 26–28]. Mechanical stylus profilometry is widely used for surface form metrology of micro-optics including surface profile and sag height [9]. The stylus tip is typically made up of diamond in conical shape. Lateral resolution of the mechanical stylus profilometric system depends on the radius of stylus tip and resolution of lateral scanning stage. Normally, the radius of conical diamond stylus tip ranges from 1 to 10 μm and in most of cases the lateral resolution is determined by the radius of stylus tip. The vertical resolution of stylus profilometer is determined by detection of vertical displacement of stylus using a dedicated sensor, ranges from sub nanometer to several nanometers. The disadvantage of mechanical stylus profilometer is that it gives the lens profile only along single cross section [9]. Atomic force microscopic (AFM) technique provides two dimensional stylus scanning (i.e. 3D surface profile) but it is time consuming and has limited scanning range along all the three axes [22]. Both vertical and lateral resolution of the order of nanometer can be achieved [9]. AFM can scan maximum height of the order of 10–20 μm and maximum scanning area is approximately limited to 150 μm $\times$ 150 μm. Scanning electron microscope (SEM) exhibits excellent imaging capabilities and resolution buta 3-D profile of micro-lens can only be obtained by using unreliable stereoscopic techniques. In addition sub-surface structure cannot be recorded with this method [22]. The attained SEM resolution can achieve less than 1 nm [29]. Optical coherence tomography (OCT) techniques have been used for the characterization of micro-optical component and lenses [24, 25]. It is a non-contact and non-invasive cross-sectional depth imaging technique. The resolution of OCT as it is highly related to light source wavelength is limited, with the highest imaging resolution achievable of 1 μm and it is inversely proportional to penetration depth of the light source [29]. Full-field swept source optical coherence tomography (FF-SS-OCT) have been used for 3D micro-lens profile measurement in which a super-luminescent diode (SLD) is used as a low coherence light source together with an acousto-optic tunable filter (AOTF) and an electronically controlled frequency tuning device [25]. The OCT method has been employed for the measurement of graded refractive index profile of crystalline lens of fisheye [24]. The main drawback is that OCT systems are expensive because of the complexity of used equipments such as AOTF and radio frequency (RF) generator used in SS-OCT and scanning device for mirror in reference arm in time domain (TD) OCT. Laser scanning confocal microscopy (LSCM) method has been used for surface profile and focal length measurement of micro-lens array [14]. In LSCM focused light spot is moved in X-Y plane normal to optic axis and the test surface is scanned in axial direction by using a piezo-electric transducer. The surface three dimensional data is obtained by the position of X-Y stage and Z piezo scanner [9]. This method has the disadvantage that one has to scan in all the three directions for evaluation of the 3D surface profile [9, 14]. The technique should be capable of providing full field information about micro-optics (shape, size and other parameters) in a non contact/non-invasive way. The interferometric techniques are also noncontact type, non invasive and provide full field information about the shape of the optical components. The conventional interferometric technique such as Twyman-Green interferometry, Mach-Zehnder interferometry, and white light interferometry are available for testing of micro-optics [15–19]. Twyman-Green and Mach-Zehnder interferometers require complex manipulation and optimization procedure to achieve precision in the measurement. That is why it is difficult to make the process of measurement automated, especially for the entire micro-lenslet array. Fully automated white light interferometers (WLIs) are commercially available with measurement capability over a 200 mm × 200 mm area. But WLIs are not suited well to measure entire lens profile and yields accurate information only for the vertex of micro-lens. Very high precision measurement, with sub nanometer resolution, can be achieved using WLI in phase shifting mode in which a narrowband pass filter is required to produce nearly monochromatic light. However, when WLIs in phase shifting interferometric (PSI) mode is used, total depth measurement is limited [17]. The entire lens surface can be measured for flat or small numerical aperture (NA) lenses, but generally vertex portion of lens can be measured with PSI mode. WLIs in PSI mode are useful for measurement of the radius of curvature (*ROC*) of best fit sphere [17]. WLI can also be applied in vertical scanning interferometry (VSI) mode to evaluate the lens height measurement. Measurement in VSI mode can be performed without using narrow band filter in white light. Using VSI mode of WLI, information at the edge of micro-lenses is lost due to steep profile. Height information at the vertex and surrounding substrate can be used to determine lens height [17, 28].

Twyman-Green interferometer is considered an accurate tool for measurement of shape of spherical and weakly aspherical lenses. A phase shifting device (through a piezoelectric transducer device) is used to obtain phase shifted interferograms and from the phase shifted interferograms the reflected wavefront from the micro-lens is evaluated, from which the lens profile and *ROC* of the lens are calculated. Generally, the *ROC* using Twyman-Green interferometer is calculated by fringe analysis with lens in cat's eye configuration. In this case the focal point of the microscope objective coincides with the vertex of micro-lens and fringes are straight lines. Thereafter, the lens is mechanically translated until the fringes disappear, i.e., this is the case when the lens *ROC* matches with the *ROC* of the wavefront illuminating the lens. The vertical movement of the micro-lens from cat's eye position to the position where the fringes disappear is the micro-lens radius of curvature. The radius of curvature measurement using this technique depends on the precision of mechanical stage and remains in the micrometer precision range [17, 28]. Direct analysis of optical performance of micro-lenses can be also performed with Mach-Zehnder interferometer. For example the aberration of micro-lenses can be determined using Mach-Zehnder interferometer in transmission mode. Mach-Zehnder interferometer proved to be useful for the characterization of cylindrical micro-lenses [17, 28]. The interferometric techniques (Twyman-Green interferometer, Mach-Zehnder interferometer and white light interferometer) consequently require significant experimental efforts with respect to phase measurement (such as phase shifting techniques) and are thus time consuming [16–18, 23, 26]. White light interferometers are not suited to measure entire lens profile and yield only to accurate information for the micro-lens vertex [17]. Digital holography (DH) overcomes the above discussed problems. Digital holography allows one to extract both the amplitude and phase information of a wavefront transmitted through a transparent object (micro-lenslet array) from a single digitally recorded hologram by use of numerical methods [28, 30–33]. Due to numerical reconstruction, the complex object wavefront at different distances can be obtained. Digital holography provides axial resolution at nanometer scale while lateral resolution is limited by diffraction and sensor size [32, 33]. Charriere et al. [28] with numerical phase compensation proposed the characterization of micro-lenslet array by digital holographic microscopy. In the experimental set up presence of microscope objective in object arm changes the divergence of object wave and causes wavefront aberration between object wave and plane reference wave [28]. In numerical phase compensation method, one needs to compute the phase mask and multiply it to the reconstructed object wave field in image plane. The basis of computation of phase mask is the precise measurement of optical setup parameters. For the computation of correct phase mask to remove the wavefront aberration, iterative adjustments of parameters are needed. Ferraro et al. also proposed a method in which the phase mask was computed in flat portion of specimen of the recorded hologram for aberration compensation [34]. Numerical phase compensation method makes the reconstruction process complex and time consuming by iterative adjustment of parameters and extrapolation of fitted polynomial in different area of recorded hologram.

In this paper, a Mach-Zehnder based digital holographic interferometric microscopic (DHIM) system is discussed for the testing of refractive micro-lenslet array. The advantage of using the DHIM is that the distortions due to aberrations in the optical system are avoided by the interferometric comparison of reconstructed phase with and without the micro-lenslet array [35]. From the experimental point of view, first a digital hologram is recorded in the absence of the micro-lenslet array which is used as a reference hologram. A second hologram is recorded in the presence of the micro-lenslet array. The presence of the transparent micro-lenslet array will induce a phase change in the transmitted laser light. Complex amplitude of object wavefront in presence and absence of micro-lens array is then reconstructed by using Fresnel reconstruction method [33]. From the reconstructed complex amplitude, one can evaluate the phase of object wave in presence and absence of the micro-lenslet array. The phase difference between the two states of object wave will provide the information about optical path length change occurring between two states. By knowing the value of the micro-lenslet array material refractive index of and that of air, the surface profile of the micro-lenslet array is calculated. From the experimentally calculated value of sag height (h) and diameter (D) of a micro-lens in micro-lenslet array, the ROC and focal length (f) of the lens are calculated. Experiments were conducted on two micro-lenslet arrays of different specification supplied by THORLABS.

2.2 Experimental Setup and Theory

Figure 2.1a shows the schematic of experimental setup of digital holographic micro-scope. The Experimental assembly is based on Mach-Zehnder interferometer. A 5 mW He-Ne laser (Make—Melles Griot, $\lambda = 632.8$ nm) is used as a light source. The light from the laser source is divided into two beams using a beam splitter BS_1. One of the beams acts as a reference beam and the other acts as an object beam. Light in the two arms of the interferometers are expanded and collimated by a spatial fil-ter (SF) assembly [Make—Newport Corp.] and a collimating lens (CL). The light in one of the arms is passed through the object under test (micro-lenslet array). In the experiment, refractive, round and plano-convex micro-lenslet array supplied by THORLABS with micro-lens sag height ($h_l = 0.87$ μm), diameter ($D_l = 146$ μm), radius of curvature ($ROC_l = 3.063$ mm), and focal length ($f_l = 6.7$ mm) was used. The object under test was mounted on a 2D translation stage in the object arm of the Mach-Zehnder interferometer and a microscope objective (20 X, NA$=0.40$) was used to increase the lateral resolution of the digital holographic microscopic system. A similar microscope objective (20 X, NA$=0.40$) was also used in the reference arm of the interferometer to match the curvature in the object and reference wavefront. A minute angle was introduced in the reference beam to make the off axis holo-graphic system. The microscope objectives in both arms and beam splitter (BS_2) were adjusted in such a way that the interference fringes are straight; this avoids the need to perform any digital correction due to spherical aberration introduced by the microscope objectives [35]. ND filters are used in the reference arm and object arm to adjust the intensity for recording good contrast fringes in the hologram. All the opti-cal components (mirrors, lenses, beam splitters, ND filters) used in the experiments are supplied by Melles Griot, Netherland.

The object beam interferes with the reference beam at the hologram plane (CCD Plane). The hologram with intensity [33]

$$H(X, Y) = |R|^2 + |O|^2 + R^*O + RO^* \qquad (2.1)$$

is recorded by a CCD sensor (Make—Lumenera Corporation, Model—Infinity3-1M). In (2.1), R is the reference wavefront and O is the object wavefront, and * denote the complex conjugate. The pixel size on CCD sensor is 6.45 μm × 6.45 μm. Total numbers of pixels are 1392 × 1040 and the sensor chip dimension is 2/3″. The dynamic range of the CCD sensor is 8-bit. A 64 bit Intel (R) Core (TM) i5 microprocessor and CPU clock rate of 3.2 GHz computer was used to process the data. The digital hologram is stored in computer for further processing. In order to reconstruct the digital hologram, a digital reference wave R_D is used to reconstruct the digital transmitted wavefront $O(m, n)$, and is given by [33]

$$O = R_D H = R_D|R|^2 + R_D|O|^2 + R_D R^*O + R_D RO^* \qquad (2.2)$$

(a)

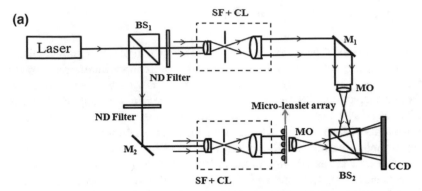

BS: Beam Splitter, ND: Neutral Density Filter, SF: Spatial Filter,
CL: Collimator M: Mirror, MO: Microscope Objective, CCD: CCD Sensor

(b)

Fig. 2.1 **a** Schematic of Mach-Zehnder interferometer based Digital Holographic Interferometric Microscope (DHIM) used for the characterization of micro-lenslet array. **b** Image of Digital Holographic Interferometric Microscope used for the testing of refractive micro-lenslet array

The first two terms of right hand side of (2.2), are the dc terms, which correspond to the zero order diffraction and the third term is the virtual image. The fourth term is the real image. To avoid the overlap between these three components (the dc term, virtual image and real image) during reconstruction, the hologram is recorded in off-axis geometry. For this purpose, the angle of reference beam (θ) with normal to CCD plane is adjusted such that θ is sufficiently large enough to ensure separation between real and virtual images in reconstruction plane. However, the angle (θ) should be small enough so that spatial frequency of micro interference pattern does not exceed than the resolving power of CCD sensor [32]. Figure 2.1b shows the photograph of the digital holographic interferometric microscope used for the characterization of refractive micro-lenslet array.

Hologram Reconstruction
In digital holography reconstruction of object wavefront is carried out by numerical methods by simulating the diffraction of reference wave at the microstructure of recorded digital hologram using scalar diffraction theory. The most commonly used

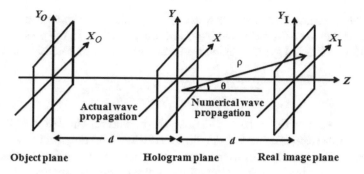

Fig. 2.2 Cartesian co-ordinate system used in the Fresnel Reconstruction method

numerical reconstruction methods are Fresnel reconstruction method, convolution method, and phase shifting method. The diffraction of reconstructing wave at the digital hologram is described by the Fresnel-Kirchhoff integral [33, 36, 37]. Here (X_O, Y_O), (X, Y), and (X_I, Y_I) are the Cartesian co-ordinate system of the object, hologram and image planes respectively. The schematic of Cartesian co-ordinate system used for the recording and reconstruction of digital hologram is shown in Fig. 2.2.

By using the Fresnel-Kirchoff integral [33, 36, 37], the complex amplitude of object wave in image plane can be calculated from

$$O(X_I, Y_I) = \frac{i}{\lambda} \int_{-\infty}^{\infty} \int_{-\infty}^{\infty} H(X, Y) R_D(X, Y) \frac{\exp(-i\frac{2\pi}{\lambda}\rho)}{\rho} \times \left(\frac{1}{2} + \frac{1}{2} \cos\theta \right) dX dY$$

(2.3)

with

$$\rho = \sqrt{(X_I - X)^2 + (Y_I - Y)^2 + d^2}$$

(2.4)

where $H(X, Y)$ is the intensity distribution in hologram plane, λ is the wavelength of light used in the experiment, ρ is the distance between a point in the hologram plane and a point in the image plane and R_D is the reference wave. The angle 'θ' is the angle between the direction of propagation of light (i.e. Z axis) and the line joining the two points, one in hologram plane and other in image plane. The angle 'θ' is defined in Fig. 2.2. In practical situation in digital holography, θ is very small, so inclination factor can be set $\left(\frac{1}{2} + \frac{1}{2} \cos\theta \right) = 1$. The diffraction is calculated at a distance 'd' behind the hologram plane, where 'd' is distance between the hologram plane and image plane meaning that it reconstructs the complex amplitude of the hologram function $H(X, Y)$ in the plane of the real image. Equation (2.3) is the basis for numerical hologram reconstruction. The Fresnel reconstruction method is based on the initial assumption that the distance 'd' between the hologram plane and reconstruction (image) plane is much larger than the maximum dimension of the sensor chip (because X, Y values as well as X_I, Y_I values, are small compared to the distance 'd' between the hologram plane and image plane). In the present case, for the validity of Fresnel approximation the following condition should be satisfied

$$d^3 \gg \frac{1}{8\lambda}[(X_I - X)^2 + (Y_I - Y)^2]_{\text{max}}^2 \tag{2.5}$$

where $[(X_I - X)^2 + (Y_I - Y)^2]_{\text{max}}^2$ denotes the maximum dimension of the sensor chip. With this restriction on 'd', the diffracted wave field remains in the near field or Fresnel diffraction region. Complex amplitude of object wavefront in image plane can be calculated from Fresnel-Kirchoff integral with Fresnel approximation and is written as [33, 36]

$$O(X_I, Y_I) = \frac{i}{\lambda d} \exp\left(-i\frac{2\pi}{\lambda}d\right) \exp\left(-i\frac{\pi}{\lambda d}\left(X_I^2 + Y_I^2\right)\right)$$

$$\times \int\limits_{-\infty}^{\infty} \int\limits_{-\infty}^{\infty} H(X, Y)R_D(X, Y)\exp\left[-i\frac{\pi}{\lambda d}\left(X^2 + Y^2\right)\right]\exp\left[i\frac{2\pi}{\lambda d}(XX_I + YY_I)\right]dX dY \tag{2.6}$$

$O(X_I, Y_I)$ can be digitized if the recorded intensity in hologram plane $H(X, Y)$ is sampled on a rectangular raster of $M \times N$ matrix points, with steps ΔX and ΔY along the co-ordinates. X_I and Y_I are replaced by $m\Delta X_I$ and $n\Delta Y_I$. With these discrete values, (2.6) can be expressed as [33, 36, 38]

$$O(mX_I, nY_I) = \frac{i}{\lambda d} \exp\left(-i\frac{2\pi}{\lambda}d\right) \exp\left[-i\frac{\pi}{\lambda d}\left(m^2\Delta X_I^2 + n^2\Delta Y_I^2\right)\right]$$

$$\times \sum_{p=0}^{M-1}\sum_{q=0}^{N-1} R_D(p, q)H(p, q)\exp\left[-i\frac{\pi}{\lambda d}\left(p^2\Delta X^2 + q^2\Delta Y^2\right)\right]$$

$$\times \exp\left[i\frac{2\pi}{\lambda d}(p\Delta Xm\Delta X_I + q\Delta Yn\Delta Y_I)\right] \tag{2.7}$$

where $O(mX_I, nY_I)$ is a $M \times N$ points matrix describing the object wavefront in the reconstruction plane, m and n are the integers ($m = 0, 1, 2, 3, ..., M - 1$; and $n = 0, 1, 2, 3, ..., N - 1$. In practice $M \times N$ corresponds to the number of pixels in the CCD sensor and ΔX and ΔY are its pixel size. $R_D(p, q)$ is the digitized reference wave; $H(p, q)$ is the recorded digital hologram; λ is the wavelength, and d is the reconstruction distance respectively. ΔX_I and ΔY_I are the pixel sizes in the reconstructed image. The relationship between pixel sizes of hologram plane and image plane are given below

$$\Delta X_I = \frac{\lambda d}{M\Delta X}; \Delta Y_I = \frac{\lambda d}{N\Delta Y}; \tag{2.8}$$

which are in compliance with the sampling theorem, which requires that the angle between the object beam and the reference beam at any point of the CCD sensor

be limited in such a way that the micro-interference fringe spacing should be larger
than double the size of pixels [36, 38]. By using the above relationship, (2.7) can be
written as

$$
O(m\Delta X_I, n\Delta Y_I) = \frac{i}{\lambda d}\exp\left(-i\frac{2\pi}{\lambda}d\right)\exp[-i\,\pi\lambda\,d(\frac{m^2}{M^2\Delta X^2}
$$

$$
+ \frac{n^2}{N^2\Delta Y^2})]\sum_{p=0}^{M-1}\sum_{q=0}^{N-1}R_D(p,q)H(p,q)
$$

$$
\times \exp[-i\frac{\pi}{\lambda d}(p^2\Delta X^2 + q^2\Delta Y^2)] \times \exp[i2\pi(\frac{pm}{M} + \frac{qn}{N})]
$$

$$
(2.9)
$$

Equation (2.9) is the discrete Fresnel transform. The matrix $O(m\Delta X_I, n\Delta Y_I)$
is calculated by multiplying $R_D(p, q)$ with $H(p, q)$ and a quadratic phase factor
$\exp[i\frac{\pi}{\lambda d}(p^2\Delta X^2 + q^2\Delta Y^2)]$ and applying an inverse discrete Fourier transform to
the product. The calculation is done most effectively using the fast Fourier transform
(FFT) [36, 38, 39].

$$
O(m\Delta X_I, n\Delta Y_I) = \frac{i}{\lambda d}\exp[-i\,\pi\lambda\,d(\frac{m^2}{M^2\Delta X^2} + \frac{n^2}{N^2\Delta Y^2})] \times FFT\{R_D(p,q)H(p,q)
$$

$$
\times \exp[-i\frac{\pi}{\lambda d}(p^2\Delta X^2 + q^2\Delta Y^2)]\}
$$

$$
(2.10)
$$

If one assumes that reference wave is a plane wave of wavelength λ, R_D can be
expressed as:

$$
R_D = \exp\left[i\frac{2\pi}{\lambda}(k_x p\Delta x + k_y q\Delta y)\right]
$$

$$
(2.11)
$$

where k_x and k_y are the components of wave vector.

To remove the dc term and -1 order term, the recorded raw digital hologram
$H(p, q)$ is filtered in Fourier domain [40, 41]. For this purpose, first we perform the
Fourier transform on $H(p, q)$. The Fourier spectrum of $H(p, q)$ gives the zero order
term (dc term), $+1$ order and -1 order. Now to remove the dc term and -1 order
term, a band pass filter is applied on the $+1$ order term. The inverse Fourier transform of the selected spectrum ($+1$ order) provides the complex amplitude containing
information about the object wavefront $O(m\Delta X_I, n\Delta Y_I, Z = 0)$. The complex
amplitude $O(m\Delta X_I, n\Delta Y_I, d)$ at a distance d parallel to CCD plane is computed
from the filtered spectrum by using Fresnel reconstruction method given by (2.10).
Numerical reconstruction of recorded digital hologram $H(p, q)$ yields the complex
amplitude of object wavefront. Once the complex amplitude of object wavefront is
calculated, the intensity and phase of the object can be calculated. The intensity of
object wavefront is calculated as [33]

$$
I(m\Delta X_I, n\Delta Y_I) = |O(m\Delta X_I, n\Delta Y_I)|^2
$$

$$
(2.12)
$$

The phase is calculated as

$$\phi(m\Delta X_I, n\Delta Y_I) = \arctan \frac{\mathrm{Im}[O(m\Delta X_I, n\Delta Y_I)]}{\mathrm{Re}[O(m\Delta X_I, n\Delta Y_I)]} \tag{2.13}$$

where Re and Im denote real and imaginary part of a complex function.

First, a digital hologram $H_1(p, q)$ is recorded in the absence of micro-lenslet array. Second digital hologram $H_2(p, q)$ is recorded in the presence of micro-lenslet array. The complex amplitude of object wavefronts $O_1(m\Delta X_I, n\Delta Y_I)$ in the absence of micro-lenslet array and $O_2(m\Delta X_I, n\Delta Y_I)$ in the presence of micro-lenslet array are reconstructed from the recorded digital holograms $H_1(p, q)$ and $H_2(p, q)$ respectively, using (2.10). The phase of object wavefronts in absence and presence of micro-lenslet array are evaluated individually from complex amplitude of object wave front $O_1(m\Delta X_I, n\Delta Y_I)$ and $O_2(m\Delta X_I, n\Delta Y_I)$ respectively. The phase of object wavefronts $\phi_1(m\Delta X_I, n\Delta Y_I)$ in absence of micro-lenslet array and $\phi_2(m\Delta X_I, n\Delta Y_I)$ in presence of micro-lenslet array can be written as

$$\phi_1(m\Delta X_I, n\Delta Y_I) = \arctan \frac{\mathrm{Im}[O_1(m\Delta X_I, n\Delta Y_I)]}{\mathrm{Re}[O_1(m\Delta X_I, n\Delta Y_I)]} \tag{2.13a}$$

$$\phi_2(m\Delta X_I, n\Delta Y_I) = \arctan \frac{\mathrm{Im}[O_2(m\Delta X_I, n\Delta Y_I)]}{\mathrm{Re}[O_2(m\Delta X_I, n\Delta Y_I)]} \tag{2.13b}$$

The phase takes values between $-\pi$ and π, the principal value of arctan function. The interference phase, which is phase difference between the phase in presence of micro-lenslet array and absence of micro-lenslet array, is calculated by modulo 2π subtraction [33, 36, 38]

$$\Delta\phi(m, n) = \begin{cases} \phi_2(m, n) - \phi_1(m, n) & if\ \phi_2(m, n) \geq \phi_1(m, n) \\ \phi_2(m, n) - \phi_1(m, n) + 2\pi & if\ \phi_2(m, n) < \phi_1(m, n) \end{cases} \tag{2.14}$$

The modulo 2π phase difference map of micro-lenslet array to ambient air is unwrapped using Goldstein phase unwrapping method to remove the 2π phase discontinuity [42].

2.3 Experimental Results

2.3.1 Reconstruction of USAF Resolution Test Chart

Initially the experiment was carried out on USAF resolution test chart. Figure 2.3a shows the recorded hologram of the USAF Resolution test target. A Fourier spectrum of the hologram gives the virtual image, real image and dc term (zero order diffraction). Figure 2.3b shows the Fourier spectrum of the recorded digital hologram of USAF resolution test chart. A band pass filter centered at frequency co-ordinate of +1 order with spatial frequency bandwidth of filter $2R_u$ is applied to remove the dc

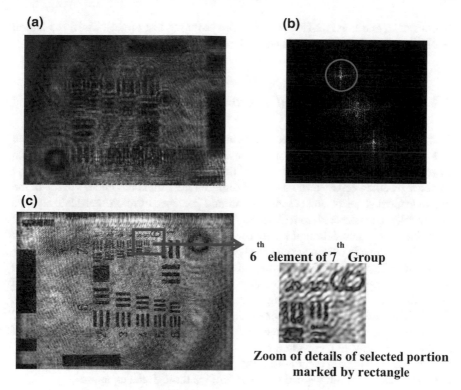

(a)

(b)

(c)

6th **element of 7th Group**

**Zoom of details of selected portion
marked by rectangle**

Fig. 2.3 **a** Hologram of USAF resolution chart. **b** Fourier spectrum of hologram. **c** Reconstructed intensity image of USAF resolution chart

term (zero order) and real image (-1 order). The bandwidth of the filter is chosen so that it does not degrade the resolution in reconstructed image. The theoretical bandwidth of filter depends on the optical magnification and numerical aperture of the microscope objective. The lateral resolution limit of microscope objective is given by Abbe criterion ($0.61\lambda/NA = 0.965$ μm for $NA = 0.40$ and $\lambda = 632.8$ nm). With optical magnification (20 x), n object of size 0.965 μm (i.e. lateral resolution limit of microscope objective) in the object plane will be imaged by 19.3 μm ($\rho_x = 20 \times 0.965$ μm $= 19.3$ μm) in the CCD plane. Thus, the total spatial frequency bandwidth ($2R_u$) of band pass filter is chosen at $2R_u = 1/\rho_x = 51.81$ line pair/mm. Red circular line in Fig. 2.3b shows the $+1$ order in Fourier spectrum that is to be filtered. Inverse Fourier transform of filtered spectrum gives complex amplitude of object wavefront at CCD plane. Complex amplitude of object wavefront in reconstruction plane is calculated by solving (2.10). The intensity image of the USAF resolution test chart is evaluated using (2.12) and is shown in Fig. 2.3c.

From the intensity image of resolution test target, it is clear that the small details of 6th element of 7th group is resolved and it corresponds to 228.1 lines pair/mm (i.e. width of one line is 2.19 μm in USAF resolution test target). Thus, the lateral resolution of the DHIM set up is better than 2.19 μm.

2.3.2 Results of Testing of Micro-lenslet Array

First, a digital hologram of ambient air (in absence of micro-lenslet array) is recorded as a reference hologram. Now, micro-lenslet array is mounted on 2D translational stage and inserted in the object arm of the Mach–Zehnder interferometer. In the presence of micro-lenses array a second digital hologram is recorded. Phases in the two individual states of object (ambient air and presence of micro-lenslet array) are numerically reconstructed from (2.13a) and (2.13b) respectively. Figure 2.4 shows the flow chart of the calculation of height map of micro-lenslet array from the recorded digital holograms in presence and absence of micro-lenslet array. Figure 2.5a shows the modulo 2π phase difference map of micro-lenslet array and ambient air. The modulo 2π phase difference map of micro-lenslet array and ambient air is unwrapped using Goldstein phase unwrapping method to remove the 2π phase discontinuity. Figure 2.5b shows the 2D unwrapped phase difference map of micro-lenslet array and ambient air. Figure 2.5c shows the 3D unwrapped phase difference map of micro-lenslet array and ambient air.

Now the optical path length difference $[\Delta n \times h(x, y)]$ can be connected to experimentally calculated unwrapped phase difference through the equation

$$\Delta\phi(x, y) = \frac{2\pi}{\lambda}\Delta n \times h(x, y) \qquad (2.15)$$

where Δn is the refractive index change $(n_s - n_0)$, n_s is refractive index of micro-lenslet material and n_0 is the ambient air refractive index, and $h(x, y)$ is the distance travelled by laser light ($\lambda = 632.8$ nm) through the micro-lenslet array.

As the micro-lenslet array is made up of fused silica its refractive index can be considered homogeneous and equal to that of fused silica which is $n_s = 1.457$ at wavelength 632.8 nm. The refractive index of air can be taken as $n_0 = 1$. The height distribution of the micro-lenses in micro-lenslet array can be evaluated from (2.15). Figure 2.6a shows the 3D height map of the micro-lenses. Figure 2.6b shows the height profile of micro-lenses along the line AB as marked in Fig. 2.5b.

The experimentally evaluated value of diameter of a lens in the micro-lenslet array is $D = 146$ μm and maximal height (sag) is $h = 0.81$ μm. From the experimentally evaluated values of diameter (D) and sag height (h), the radius of curvature (ROC) and focal length (f) of the same lens in micro-lenslet array are computed according to equations [35, 43]

$$ROC = \frac{h}{2} + \frac{D^2}{8h} \qquad (2.16)$$

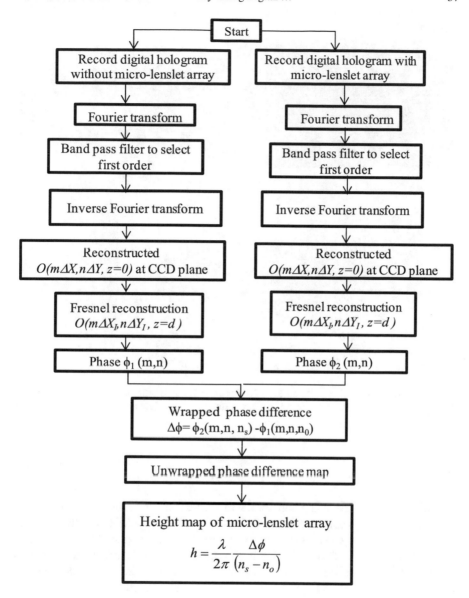

Fig. 2.4 Flow chart of calculation of height map of micro-lenslet array

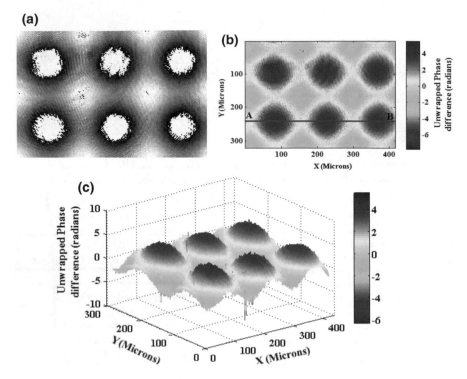

Fig. 2.5 **a** Modulo 2π phase difference, **b** 2D unwrapped phase difference map of micro-lenslet array and ambient air, and **c** 3D unwrapped phase difference map of micro-lenslet array and ambient air

$$\frac{1}{f} = \frac{n-1}{ROC} \tag{2.17}$$

The experimentally computed values of parameters of a micro-lens in micro-lenslet array using DHIM are $D = 145.9$ μm, $h = 0.81$ μm, $ROC = 3.289$ mm and $f = 7.19$ mm. These values agree very well with the ones supplied by manufacturer ($D = 146$ μm, $h = 0.87$ μm, $ROC = 3.063$ mm and $f = 6.7$ mm). Table 2.1 shows the comparison of optical parameters of fused silica micro-lens in the micro-lenslet array using DHIM and data provided by supplier (THORLABS).

Surface height map measurement was also carried out with commercially available Coherence Correlation Interferometer (CCI) from Taylor Hobson Ltd. UK with 0.1 Å axial resolution. Figure 2.7a shows the two dimensional height map of micro-lenses in the micro-lenslet array measured by CCI and Fig. 2.7b shows the height profile of micro-lenses along the line AB as marked by dotted line in Fig. 2.7a.

Height profile of micro-lenses measured by DHIM is compared with commercially available Coherence Correlation Interferometer (Manufacturer: Taylor Hobson Ltd. UK, Axial resolution 0.1 Å). Figure 2.8 shows the comparison of height profiles of micro-lenses array obtained by DHIM and Coherence Correlation Interferometer

Fig. 2.6 **a** 3D height map of
micro-lenslet array. **b** Height
profile of micro-lenses along
the line AB as marked in
Fig. 2.5b

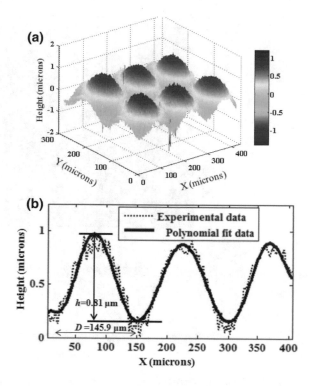

Table 2.1 Comparison between the optical parameters obtained by DHIM and data provided by
supplier of a micro-lens in first micro-lenslet array

Micro-lens parameters	DHIM calculated value	Data provided by supplier (THORLAB)	Deviation (%)
Diameter (*D*)/pitch	145.9 μm	146 μm	0.68
Sag height (*h*)	0.81 μm	0.87 μm	6.89
Radius of Curvature (*ROC*)	3.289 mm	3.063 mm	7.37
Focal length (*f*)	7.19 mm	6.7 mm	7.37

(CCI). Root mean square error (RSME) between the measurement done by DHIM
and CCI is 0.12%.

The experiments were also conducted on the refractive, square and plano-
convex micro-lenslet array supplied by THORLABS with micro-lens sag height (*h*
= 1.31 μm), pitch (300 μm), radius of curvature (*ROC* = 8.6 mm), and focal length
(*f* = 18.6 mm). In the experiment, 10 x microscope objectives were used in the object
and reference arm of the Mach-Zehnder based DHIM system to increase the field of
view. Figure 2.9a shows the modulo 2π phase difference map of micro-lenslet array
and ambient air. This modulo 2π phase difference map of micro-lenslet array to that
of ambient air was unwrapped using Goldstein phase unwrapping method to remove

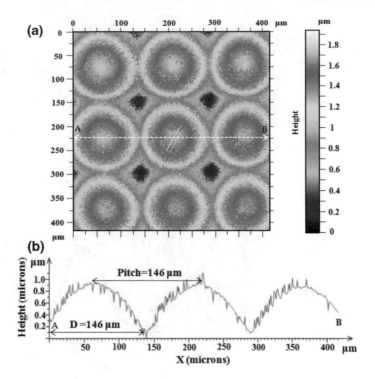

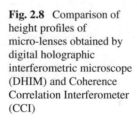

Fig. 2.7 **a** 2D height map of micro-lenses in micro-lenslet array using coherence correlation interferometer (CCI). **b** Height profile of micro-lenses measured by coherence correlation interferometer

Fig. 2.8 Comparison of height profiles of micro-lenses obtained by digital holographic interferometric microscope (DHIM) and Coherence Correlation Interferometer (CCI)

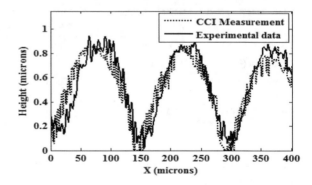

the 2π phase discontinuity. Figure 2.9b shows the 2D unwrapped phase difference map of micro-lenslet array and ambient air and Fig. 2.9c shows the 3D unwrapped phase difference map of micro-lenslet array to that of ambient air.

Figure 2.10a shows the 3D height map of the micro-lenses and Fig. 2.10b shows the height profile of micro-lenses along the line AB as marked in Fig. 2.9b.

The experimentally micro-lens computed parameters values in the micro-lenslet array, using DHIM, are Pitch = 302 μm, h = 1.236 μm, ROC = 8.79 mm and

Fig. 2.9 **a** Modulo 2π phase difference, **b** 2D unwrapped phase difference map of micro-lenslet array and ambient air, and **c** 3D unwrapped phase difference map of micro-lenslet array and ambient air

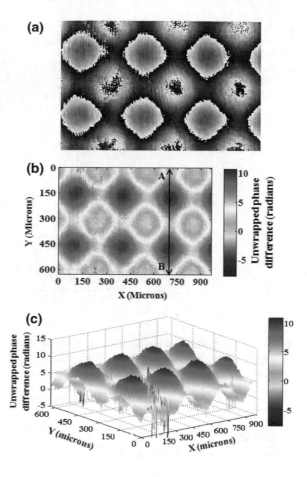

$f = 19.24$ mm. These values agree very well with the ones supplied by manufacturer (pitch $= 300$ μm, $h = 1.31$ μm, $ROC = 8.6$ mm and $f = 18.6$ mm). Table 2.2 shows the comparison of optical parameters of fused silica micro-lens in the micro-lenslet array using DHIM and data provided by supplier (THORLABS).

2.4   Discussion and Conclusions

In this chapter, the potential of Mach-Zehnder based off- axis digital holographic interferometric microscope (DHIM) as a metrological tool for the characterization of the micro-lenslet array was demonstrated. In USAF resolution test chart the width of smallest bar is 2.19 μm (i.e. 6th element of 7th group). A lateral resolution, better than 2.19 μm, can be attained using DHIM. The theoretical resolution limit is 0.965 μm. It is also demonstrated that the bandwidth of the filter is chosen so that it does not

Fig. 2.10 **a** 3D height map of micro-lenses in the micro-lenslet array. **b** Height profile of micro-lenses along the line AB as marked in Fig. 2.9b

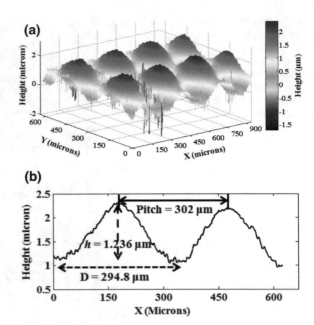

Table 2.2 Comparison between the optical parameters obtained by DHIM and data provided by supplier of a micro-lens in second micro-lenslet array

Micro-lens parameters	DHIM calculated value	Data provided by supplier (THORLAB)	Deviation (%)
Pitch	302 μm	300 μm	0.66
Sag height (h)	1.236 μm	1.31 μm	5.64
Radius of Curvature (ROC)	8.79 mm	8.6 mm	2.20
Focal length (f)	19.24 mm	18.6 mm	3.44

degrade the reconstructed image resolution. A smaller size of spatial frequency bandwidth ($2R_u$) of band pass filter means that most of the high frequencies are filtered out, so that the loss of details in the reconstructed images is large. However a large size of spatial frequency bandwidth ($2R_u$) of band pass filter will add the effect of the zero-order spectrum and gives rise to erroneous measurement. The theoretical filter bandwidth depends on the optical magnification and numerical aperture of the microscope objective. The total spatial frequency bandwidth ($2R_u$) of band pass filter is chosen as 51.81 line pair/mm. In the experiments two refractive, plano–convex micro-lenslet arrays of different specifications were used as test objects. The micro-lenslet arrays were supplied by THORLABS (USA). For the first micro-lenslet array, the measured parameters values of a lens in the micro-lenslet array attained by DHIM were—diameter $D = 145.9$ μm, sag height $h = 0.81$ μm, radius of curvature $ROC = 3.289$ mm, and focal length $f = 7.19$ mm. These values agree pretty well with the ones provided by the supplier ($D = 146$ μm, $h = 0.87$ μm, $ROC = 3.063$ mm and $f = 6.7$ mm). The corresponding deviations are 0.07% for diameter, 6.89% for sag height, 7.37% for radius of curvature and 7.37% for focal length. The height profile of micro-lenses measured by DHIM has been compared with commercially available Coherence Correlation Interferometer (Manufacturer: Taylor Hobson Ltd. UK, Axial resolution 0.1 Å). Root mean square error (RSME) between the measurement done by DHIM and CCI is 0.12%. For the second microlenslet array, the measured value of parameters of a lens in micro-lenslet array from DHIM were—Pitch $= 302$ μm, sag height $h = 1.236$ μm, radius of curvature $ROC = 8.79$ mm, and focal length $f = 19.24$ mm. These values agree well with those provided by the supplier (Pitch $= 300$ μm, $h = 1.31$ μm, $ROC = 8.6$ mm and $f = 18.6$ mm). The corresponding deviations are 0.66% for pitch, 5.64% for sag height, 2.20% for radius of curvature and 3.44% for focal length.

Acknowledgements The financial assistance received from the Defence Research and Development Organization (DRDO), Ministry of Defence, Government of India, under the project entitled 'Testing of micro optics using digital holographic interferometry' under FA sanction No. ERIP/ER/1300466/M/01/1556 dated 20 Nov. 2014 is highly acknowledged.

References

1. S. Sinzinger, J. Jahns, *Microoptics*. Wiley (1999)
2. R.H. Anderson, Close-up imaging of documents and displays with lens arrays. Appl. Opt. **18**(4), 477–484 (1979)
3. F.B. McCormick, F.A.P. Tooley, T.J. Cloonan, J.M. Sasian, H.S. Hinton, K.O. Mersereau, A.Y. Feldblum, Optical interconnections using microlens arrays. Opt. Quantum Electron. **24**(4), S465–S477 (1992)
4. T. Hou, C. Zheng, S. Bai, Q. Ma, D. Bridges, A. Hu, W.W. Duley, Fabrication, characterization, and applications of microlenses. Appl. Opt. **54**(24), 7366–7376 (2015)
5. SUS MicroOptics SA, Catalog (2007), pp. 1–20, http://www.amstechnologies.com/fileadmin/amsmedia/downloads/2067_SMO_catalog.pdf
6. M. Stedman, K. Lindsey, Limits of surface measurement by stylus instruments, in *1988 International Congress on Optical Science and Engineering* (International Society for Optics and Photonics, 1989), pp. 56–61
7. K.W. Lee, Y.J. Noh, Y. Arai, Y. Shimizu, W. Gao, Precision measurement of micro-lens profile by using a force-controlled diamond cutting tool on an ultra-precision lathe. Int. J. Precis. Technol. **2**(2–3), 211–225 (2011)
8. D.H. Lee, N.G. Cho, Assessment of surface profile data acquired by a stylus profilometer. Meas. Sci. Technol. **23**(10), 105601 (2012)
9. B. Xu, Z. Jia, X. Li, Y.L. Chen, Y. Shimizu, S. Ito, W. Gao, Surface form metrology of micro-optics, in *International Conference on Optics in Precision Engineering and Nanotechnology (icOPEN2013)* (International Society for Optics and Photonics, 2013), pp. 876902–876902
10. J. Aoki, W. Gao, S. Kiyono, T. Ono, A high precision AFM for nanometrology of large area micro-structured surfaces, in *Key Engineering Materials,* vol. 295. (Trans Tech Publications, 2005), pp. 65–70
11. A. Yacoot, L. Koenders, Recent developments in dimensional nanometrology using AFMs. Meas. Sci. Technol. **22**(12), 122001 (2011)

12. W. Gao, S. Goto, K. Hosobuchi, S. Ito, Y. Shimizu, A noncontact scanning electrostatic force microscope for surface profile measurement. CIRP Ann. Manuf. Technol. **61**(1), 471–474 (2012)
13. H.J. Jordan, M. Wegner, H. Tiziani, Highly accurate non-contact characterization of engineering surfaces using confocal microscopy. Meas. Sci. Technol. **9**(7), 1142 (1998)
14. H.J. Tiziani, T. Haist, S. Reuter, Optical inspection and characterization of microoptics using confocal microscopy. Opt. Lasers Eng. **36**(5), 403–415 (2001)
15. J. Schwider, O.R. Falkenstoerfer, Twyman-Green interferometer for testing microspheres. Opt. Eng. **34**(10), 2972–2975 (1995)
16. S. Reichelt, H. Zappe, Combined Twyman–Green and Mach–Zehnder interferometer for microlens testing. Appl. Opt. **44**(27), 5786–5792 (2005)
17. K.J. Weible, R. Volkel, M. Eisner, S. Hoffmann, T. Scharf, H.P. Herzig, Metrology of refractive microlens arrays, in *Photonics Europe* (International Society for Optics and Photonics, 2004), pp. 43–51
18. V. Gomez, H. Ottevaere, H. Thienpont, Mach–Zehnder interferometer for real-time in situ monitoring of refractive microlens characteristics at the fabrication level. IEEE Photonics Technol. Lett. **20**(9), 748–750 (2008)
19. H. Sickinger, O.R. Falkenstoerfer, N. Lindlein, J. Schwider, Characterization of microlenses using a phase-shifting shearing interferometer. Opt. Eng. **33**(8), 2680–2686 (1994)
20. X. Zhu, S. Hu, L. Zhao, Focal length measurement of a microlens-array by grating shearing interferometry. Appl. Opt. **53**(29), 6663–6669 (2014)
21. U.P. Kumar, N.K. Mohan, M.P. Kothiyal, Characterization of micro-lenses based on single interferogram analysis using Hilbert transformation. Opt. Commun. **284**(21), 5084–5092 (2011)
22. A.J. Krmpot, G.J. Tserevelakis, B.D. Murić, G. Filippidis, D.V. Pantelić, 3D imaging and characterization of microlenses and microlens arrays using nonlinear microscopy. J. Phys. D: Appl. Phys. **46**(19), 195101 (2013)
23. H.H. Wahba, T. Kreis, Characterization of graded index optical fibers by digital holographic interferometry. Appl. Opt. **48**(8), 1573–1582 (2009)
24. E. Acosta, L. Garner, G. Smith, D. Vazquez, Tomographic method for measurement of the gradient refractive index of the crystalline lens. I. The spherical fish lens. J. Opt. Soc. Am. A **22**(3), 424–433 (2005)
25. T. Anna, C. Shakher, D.S. Mehta, Three-dimensional shape measurement of micro-lens arrays using full-field swept-source optical coherence tomography. Opt. Lasers Eng. **48**(11), 1145–1151 (2010)
26. T. Zhang, I. Yamaguchi, Three-dimensional microscopy with phase-shifting digital holography. Opt. Lett. **23**(15), 1221–1223 (1998)
27. V. Kebbel, J. Mueller, W.P. Jueptner, Characterization of aspherical micro-optics using digital holography: improvement of accuracy, in *International Symposium on Optical Science and Technology* (International Society for Optics and Photonics, 2002), pp. 188–197
28. F. Charrière, J. Kühn, T. Colomb, F. Montfort, E. Cuche, Y. Emery, K. Weible, P. Marquet, C. Depeursinge, Characterization of microlenses by digital holographic microscopy. Appl. Opt. **45**(5), 829–835 (2006)
29. Y. Wei, C. Wu, Y. Wang, Z. Dong, Efficient shape reconstruction of microlens using optical microscopy. IEEE Trans. Ind. Electron. **62**(12), 7655–7664 (2015)
30. U. Schnars, W. Jüptner, Direct recording of holograms by a CCD target and numerical reconstruction. Appl. Opt. **33**(2), 179–181 (1994)
31. E. Cuche, P. Marquet, C. Depeursinge, Simultaneous amplitude-contrast and quantitative phase-contrast microscopy by numerical reconstruction of Fresnel off-axis holograms. Appl. Opt. **38**(34), 6994–7001 (1999)
32. E. Cuche, F. Bevilacqua, C. Depeursinge, Digital holography for quantitative phase-contrast imaging. Opt. Lett. **24**(5), 291–293 (1999)
33. U. Schnars, W. Jueptner, *Digital Holography: Digital Hologram Recording, Numerical Reconstruction and Related Techniques* (Springer, Berlin, Heidelberg, 2005)

34. P. Ferraro, S. De Nicola, A. Finizio, G. Coppola, S. Grilli, C. Magro, G. Pierattini, Compensation of the inherent wave front curvature in digital holographic coherent microscopy for quantitative phase-contrast imaging. Appl. Opt. **42**(11), 1938–1946 (2003)
35. V. Kumar, C. Shakher, Testing of micro-optics using digital holographic interferometric microscopy, in *Proceedings of the 4th International Conference on Photonics, Optics and Laser Technology (PHOTOPTICS)* (SCITEPRESS—Science and Technology Publications, Lda, 2016), pp. 142–147
36. U. Schnars, W.P. Jüptner, Digital recording and numerical reconstruction of holograms. Meas. Sci. Technol. **13**(9), R85 (2002)
37. C. Wagner, S. Seebacher, W. Osten, W. Jüptner, Digital recording and numerical reconstruction of lensless Fourier holograms in optical metrology. Appl. Opt. **38**(22), 4812–4820 (1999)
38. V. Kumar, M. Kumar, C. Shakher, Measurement of natural convective heat transfer coefficient along the surface of a heated wire using digital holographic interferometry. Appl. Opt. **53**(27), G74–G83 (2014)
39. V. Kumar, C. Shakher, Study of heat dissipation process from heat sink using lensless Fourier transform digital holographic interferometry. Appl. Opt. **54**(6), 1257–1266 (2015)
40. M. Takeda, Spatial-carrier fringe-pattern analysis and its applications to precision interferometry and profilometry: an overview. J. Sci. Ind. Metrol. **1**(2), 79–99 (1990)
41. E. Cuche, P. Marquet, C. Depeursinge, Spatial filtering for zero-order and twin-image elimination in digital off-axis holography. Appl. Opt. **39**(23), 4070–4075 (2000)
42. R.M. Goldstein, H.A. Zebker, C.L. Werner, Satellite radar interferometry: two-dimensional phase unwrapping. Radio Sci. **23**(4), 713–720 (1988)
43. J. Kühn, F. Charrière, T. Colomb, E. Cuche, Y. Emery, C. Depeursinge, Digital holographic microscopy for nanometric quality control of micro-optical components, in *Integrated Optoelectronic Devices 2007* (International Society for Optics and Photonics, 2007), pp. 64750V–64750V

Chapter 3
Estimation of the OSNR Penalty Due to In-Band Crosstalk on the Performance of Virtual Carrier-Assisted Metropolitan OFDM Systems

Bruno R. Pinheiro, João L. Rebola and Adolfo V. T. Cartaxo

Abstract The impact of the in-band crosstalk on the performance of virtual carrier (VC)-assisted direct detection (DD) multi-band orthogonal frequency division multiplexing (MB-OFDM) systems was numerically assessed via Monte-Carlo simulations, by means of a single interferer and 4-ary, 16-ary and 64-ary quadrature amplitude modulation (QAM) formats in the OFDM subcarriers. It was also investigated the influences of the virtual carrier-to-band power ratio (VBPR) and the virtual carrier-to-band gap (VBG) on the DD in-band crosstalk tolerance of the OFDM receiver. It was shown the modulation format order decrease enhances the tolerance to in-band crosstalk. When the VBG is the same for both interferer and selected signal, the interferer VBPR increase is seen to lead to lower optical signal-to-noise ratio (OSNR) penalties due to in-band crosstalk. Considering that the VCs frequencies of the selected and interferer OFDM signals are equal, the increase of the interferer VBG also gives rise to lower OSNR penalties. When the interferer and selected signals bands central frequencies are the same, the change of interferer VBG can attain 11 dB less tolerance to in-band crosstalk of the VC-assisted DD OFDM system. We also evaluate the error vector magnitude (EVM) accuracy of the in-band crosstalk tolerance of the DD OFDM receiver and our results show that the EVM estimations are inaccurate.

B. R. Pinheiro · J. L. Rebola (✉) · A. V. T. Cartaxo
Optical Communications and Photonics, Instituto de Telecomunicações, Lisbon, Portugal
e-mail: joao.rebola@iscte-iul.pt

B. R. Pinheiro
e-mail: bruno_pinheiro@iscte-iul.pt

A. V. T. Cartaxo
e-mail: adolfo.cartaxo@iscte-iul.pt

B. R. Pinheiro · J. L. Rebola · A. V. T. Cartaxo
Instituto Universitário de Lisboa (ISCTE-IUL), Lisbon, Portugal

© Springer Nature Switzerland AG 2018
P. A. Ribeiro and M. Raposo (eds.), *Optics, Photonics and Laser Technology*, Springer
Series in Optical Sciences 218, https://doi.org/10.1007/978-3-319-98548-0_3

47

3.1 Introduction

Metropolitan networks aggregate different types of traffic and provide links between access and back-bone networks. In addition, these networks need to present a fast traffic exchange as a result of the protocols employed to aggregate different traffic types. Hence, metropolitan networks must present high flexibility, dynamic reconfigurability and transparency, and should enable scalability [4]. Additionally, the lower power consumption and less space occupied by network elements have been important requirements for metro networks planning from operators viewpoint [4]. In order to respond to those requirements, hybrid optical networks, that integrate metro and access networks in the same optical network, have been selected as an attractive alternative [9]. In comparison with conventional metro and access networks, the hybrid networks should comprise less network elements, therefore, the power consumption will be potentially lower [10].

Orthogonal frequency division multiplexing (OFDM) is generally known for having a high spectral efficiency, allowing to increase the capacity and robustness of optical networks against fibre dispersion. In terms of detection schemes, there are two methods [26] (i) coherent detection, in which a local oscillator, hybrid couplers and several photodetectors at the optical receiver are used, and (ii) direct-detection where only one photodetector is required at the receiver. The coherent detection presents higher transmission performance in comparison with DD. However, it has a higher system cost and complexity. Hence, DD systems are favored for metro applications.

The OFDM technique also enables flexible bandwidth allocation, which is an important aspect for hybrid optical networks. This OFDM feature can be used to allocate bandwidth for several users with both higher energy and efficiency on the resources management. These features can be accomplished through the multi-band (MB) OFDM technique, in which several OFDM bands are simultaneously transmitted by a single wavelength [6].

The MORFEUS network, which is a virtual carrier (VC)-assisted DD MB-OFDM network [3], has been designated as a system that efficiently meets the requirements for hybrid networks aforementioned [4].

The signal-to-signal beat interference (SSBI) is an important impairment caused by photodetection [26]. The impact of the SSBI on the performance degradation is eliminated by setting a frequency gap, larger than the OFDM signal bandwidth, between the OFDM band and the VC, or by using digital signal processing algorithms that mitigate the SSBI term at the OFDM detected signal. In this work, the SSBI mitigation technique presented in [17] is implemented. The use of the SSBI mitigation technique allows to reduce the band gap between the VC and the OFDM band, and consequently, improves the system spectral efficiency [19].

The performance of metro networks can be strongly impaired by in-band crosstalk, which is the interference between signals with the same nominal wavelength. The interfering signals are due to power leakage coming from the deficient isolation of the switching devices inside the optical nodes, as for example in a reconfigurable add-drop multiplexer (ROADM) [28]. The ROADM is a network element that plays an essential role in nowadays transport networks, as they are responsible for the switching of optical signals. The optical switching, inside the ROADM, is performed by wavelength selective switches. Those devices have imperfect isolation, consequently, during the add and drop operation, in-band crosstalk signals are originated from power leakage. Then, the crosstalk signals are transmitted through the optical network, causing a strong degradation on system performance [8]. The tolerance to in-band crosstalk has been investigated in DD OFDM systems [22], where the selected OFDM signal bandwidth is equal to the band gap between the VC and the OFDM band. Nonetheless, the tolerance to in-band crosstalk of a VC-assisted DD OFDM receiver with SSBI mitigation algorithm is still to be assessed.

In this chapter, the impact of the in-band crosstalk on the performance of the VC-assisted DD MB-OFDM receiver with SSBI mitigation algorithm for different M-QAM format orders is evaluated through Monte-Carlo (MC) simulation by using direct error counting (DEC) as a bit error rate (BER) estimation method. The error vector magnitude (EVM) is also used as performance estimation method. The effect of the main parameters of the OFDM signal with VC, such as the virtual carrier-to-band power ratio (VBPR) and virtual carrier-to-band gap (VBG), on the performance of the DD OFDM receiver impaired by in-band crosstalk is also assessed using the EVM and DEC methods. The EVM accuracy is studied by comparison with the results obtained from DEC.

This chapter is organized as follows. In Sect. 3.2, some important works referring to in-band crosstalk research are reviewed and the most relevant conclusions of those studies presented. Section 3.3 describes the MORFEUS network and its simulation model. The MORFEUS network is presented in Sect. 3.3.1 and, in Sect. 3.3.2, the MC simulation is described. Numerical results are presented and discussed in Sect. 3.4. Conclusions are outlined in Sect. 3.5.

3.2 Literature Review

The study of the impact of in-band crosstalk on the performance of optical systems has been an important subject of optical communication research in the past years. In the first works related to this subject, for example [5, 15, 16], the imperfect isolation of the devices that comprise the optical nodes is identified as the origin of the crosstalk signals, and also that the in-band crosstalk is the most detrimental type of crosstalk, as it interferes the signals with same wavelength. Another main conclusion is that the main contributor of the performance degradation is the interferometric beat noise [5, 15, 16]. The interferometric beat noise is the beat noise between the primary and interfering signals that occurs at the photodetector of the receiver and becomes an relevant source of performance degradation in DD receivers [7].

Nowadays, the coherent detection is the chosen detection scheme in the transport networks, thereby, the quadrature phase-shift keying (QPSK) has became the modulation most used in these networks. In fact, the coherent detection allows the transmission of M-ary quadrature amplitude modulation (QAM), hence, the coherent detection allows the coexistence of a large variety of different modulation formats and bit rates in the optical networks [29]. This fact makes the study of the in-band crosstalk an even more relevant topic in the optical communication research, as the coherent detection enables crosstalk signals with different characteristics, such as, different modulation format order, symbol rate or bit rate, hence, leading to different impacts on the receiver performance. The impact of the in-band crosstalk on the performance of a polarization division multiplexing (PDM)-QPSK has been analytically analyzed, assuming that the in-band crosstalk has a Gaussian distribution [13]. The main conclusion was that the Gaussian model is valid for a large number of interferers. For the case of a single interfering signal, the Gaussian model overestimates the BER estimation in presence of in-band crosstalk. The weighted crosstalk has been proposed and experimentally validated as estimation method of performance penalties due to in-band crosstalk [12, 18].

Regarding the OFDM-based networks using QPSK and M-QAM modulation format, some works can be found in the literature. In [11], the performance degradation of a 112-Gb/s PDM-coherent optical-OFDM system due to the interaction between in-band crosstalk and fiber nonlinearity is estimated. The results found in [11] reveal that the fiber nonlinearity enhances the influence of the in-band crosstalk on the BER penalty of the PDM-OFDM coherent receiver. In [22, 23], it is concluded that higher modulation format orders used in the OFDM subcarriers leads to lower in-band crosstalk tolerance of the OFDM DD receiver. These conclusions are similar to the ones presented in [20, 28], in which the impact of in-band crosstalk on the performance of optical communication systems with coherent detection due to different M-QAM crosstalk signals is assessed.

Regarding the assessment tools to estimate the impact of the DD receiver in presence of in-band crosstalk, the most common and accurate method is the DEC [2, 14]. However, the estimation of very low error probabilities ($\leq 10^{-5}$) can be extremely time consuming [2]. So, several alternative performance assessment methods have been proposed and used in order to achieve the DEC accuracy with less computational effort. The moment generating function (MGF) has been proposed as a theoretical method to assess the influence of the in-band crosstalk on the performance degradation of the OFDM DD receiver [23]. It was concluded that, in absence of signal distortion due to the non-linearity of optical modulator, the MGF can predict accurately the BER of the OFDM DD receiver impaired by in-band crosstalk.

The EVM has been gaining popularity as performance assessment tool of M-QAM signals receivers, in fact, the EVM is the performance assessment tool used in [4]. The EVM is a semi-analytical method that makes use of the difference between the M-QAM symbol location in the received constellation (due to the noise detected at the receiver) and its transmitted constellation location in order to estimate the BER [25]. In absence of in-band crosstalk, the BER estimation performed by the EVM method has been reported to have a excellent agreement with the BER estimated using DEC with significantly less computation effort [2, 24]. The accuracy of the BER estimation using EVM method in presence of in-band crosstalk of coherent detection system has been assessed in [21]. The main conclusion of was that, for higher crosstalk levels, the EVM method looses accuracy on the estimation of the performance degradation of the M-QAM coherent receiver. However, to the best of our knowledge, the EVM accuracy on the BER estimation of DD OFDM receiver in presence of in-band crosstalk is still to be evaluated.

The MORFEUS network has been proposed and its operation and optimization detailed elsewhere [4]. The MORFEUS transmits a single-side band (SSB) OFDM signal, i.e. only one side of optical signal is transmitted. The transmission of SSB optical signal is known to overtake the chromatic dispersion induced power fading (CDIPF). The CDIPF arises from the conjunction of the beat between the two side-bands of optical signal that occurs at the DD receiver and the accumulated dispersion of the optical fiber [1].

The MORFEUS network allows to aggregate several OFDM signals in a single wavelength, composing the MB-OFDM signal. The MB-OFDM signal comprises, in each band, a VC and the OFDM band. The power of the VC is set based on the VBPR, which relates the average power of the VC with the average power of the corresponding OFDM band. Increasing the VBPR is a mean to overcome the SSBI, however, this solution leads to higher power consumption, thereby, higher VBPR is undesirable option to accomplish the requirements of hybrid networks. In order to overcome this disadvantage, the MORFEUS network uses a SSBI mitigation algorithm at the OFDM demodulation. Hence, the SSBI term is eliminated and, consequently, allows to reduce the VBPR. Moreover, since the SSBI term is eliminated on the OFDM detected signal, the frequency of VC can be set closely to the corresponding OFDM band, which means that the spectral efficiency is higher than in conventional OFDM networks. In these networks, in order to avoid degradation of the DD OFDM receiver performance due to SSBI, the VBG is equal or larger than the OFDM signal bandwidth [23]. The drawback of using SSBI mitigation algorithms is the increase of the receiver complexity. The implementation of SSBI mitigation algorithms is performed by digital signal processing (DSP) at the receiver that reconstructs and removes the SSBI from the photodetected signal [4].

The MORFEUS network detects each OFDM band separately using a dual band optical filter. The filter drops the desired band with its VC, enabling the demodulation of the information of the OFDM band at the DD OFDM receiver. This solution increases remarkably the spectral efficiency and reduces the required bandwidth of the DD OFDM receiver [4] in comparison with the conventional OFDM optical systems.

This work intents to make a contribution on the study of the impact of the in-band crosstalk on performance of VC-assisted DD OFDM receivers, in particular, analyzing the influence of the VBPR and VBG on the tolerance to in-band crosstalk for different modulation format orders. We also intend to investigate the accuracy of the EVM method as a figure of merit of VC-assisted DD OFDM receiver impaired with in-band crosstalk.

3.3 MORFEUS Network

The MORFEUS network and its model will be described in this section, the MB-OFDM signal main parameters detailed and finally the MC simulation layout sketched.

3.3.1 Network Model

The block diagram of the MORFEUS metro network [4] is depicted in Fig. 3.1. The MORFEUS network has a ring topology. Each network node includes a reconfigurable optical add-and-drop multiplexer (ROADM), a MORFEUS insertion block (MIB) and a MORFEUS extraction block (MEB). The MIB generates the electrical OFDM bands and VCs at the OFDM transmitter (Tx). The electrical OFDM signal is then converted to the optical domain by means of an electrical-to-optical converter (EOC), inserted in the optical network [4]. The band extraction in the MIEB is performed by a tunable optical filter (BS), which tunes to the desired OFDM signal. The tuned OFDM signal is then issued to the SSBI estimation block (SEB) and also to the PIN photodiode, to be detected. Afterwards, the estimated SSBI, obtained from the SEB, is extracted from the photodetected OFDM signal, and the signal demodulation without SSBI is carried out at the OFDM Rx.

A single OFDM band was taken into account in evaluating the impact of in-band crosstalk on the DD OFDM receiver. Under these conditions, the OFDM signal can be assumed has having only one pair OFDM band-VC, thus having the PSD spectrum as in Fig. 3.2.

The PSD spectrum shown in Fig. 3.2 is representing the OFDM signal at the transmitter output having a 11 mW average power. The OFDM band is characterized

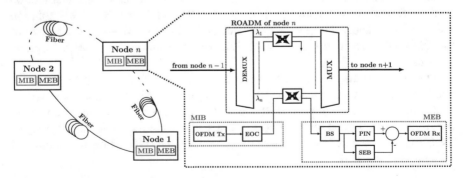

Fig. 3.1 MORFEUS metro network and respective nodes block diagram, consisting of reconfigurable optical add-and-drop multiplexer (ROADM), MORFEUS insertion block (MIB) and MORFEUS extraction block (MEB). EOC, BS and SEB stand for electrical-optical converter, band selector and SSBI estimation block, respectively

Fig. 3.2 PSD of the electrical MB-OFDM signal at the OFDM Tx output, with an average power of 11 mW, considering one pair band-VC

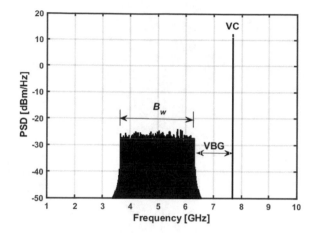

by a 2.675 GHz bandwidth, B_w, and a 5 GHz central frequency. The bandwidth B_w is stated here as N_{sc}/T_s, where N_{sc} is the number of subcarriers and T_s is the OFDM symbol duration without guard time. The frequency gap between the OFDM band and the VC is the VBG, which in Fig. 3.2 is $0.5B_w$. To maximize the SE system, the VBG is set to 20.9 MHz and the ratio between VC average power and the OFDM power of the corresponding band is VBPR.

3.3.2 Monte Carlo Simulation

This section deals with the MC simulation details and the methods used to obtain the BER.

The MORFEUS network simulation model used to assess the performance tolerance to in-band crosstalk is shown in Fig. 3.3. Basically, it includes a Tx having VC generation, a dual parallel Mach-Zehnder modulator (DP-MZM), a tunable band selector (BS), an ideal photodetector, a SEB and an OFDM receiver. One aims to estimate the BER at the OFDM receiver output. At the optical receiver input, just before the BS, amplified spontaneous emission (ASE) noise and in-band crosstalk are placed together with the OFDM signal.

The MC simulation begins with a M-QAM symbol sequence [14] representative of the electrical OFDM signal, comprising the OFDM band with a VC. The electrical-optical conversion is then carried out by the DP-MZM, which generates a single-side band OFDM optical signal by applying the electrical OFDM signal and its Hilbert transform (HT) in both arms of the modulator. Here, the OFDM signal HT is assumed ideal. The DP-MZM modulation index considered is 5%, which is the optimized value [4]. Then, ASE noise and in-band crosstalk sample functions are added to the optical OFDM signal, by considering a back-to-back configuration.

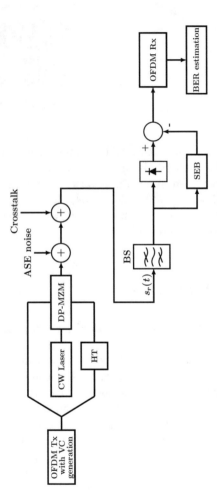

Fig. 3.3 Model of the VC-assisted DD OFDM system. The acronym DP-MZM, HT and CW stand for dual parallel Mach-Zehnder modulator, Hilbert transform and continuous wave

Fig. 3.4 The SEB model.
VCS stands for VC selector

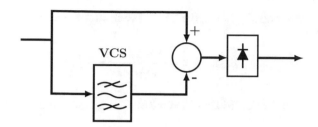

The SEB model, sketched in Fig. 3.4, is based on the SSBI mitigation technique described in [17]. At the lower SEB branch, Fig. 3.4 the VC of the selected OFDM signal is tuned by means of an ideal optical filter—a virtual carrier selector (VCS). The VC is then withdrawn from the upper branch OFDM signal and subsequently, the SSBI is predicted after the photodetection of the OFDM signal without the VC. Finally, to conclude the SSBI mitigation algorithm, the SSBI is withdrawn from the photodetected OFDM signal before reaching the OFDM receiver, as shown in Fig. 3.3. It is assumed here that both SEB branches are synchronized.

At the input of the BS, the OFDM signal, $s_r(t)$, impaired by the interferer and ASE noise can be written as

$$s_r(t) = s_0(t) + \sum_{i=1}^{N_x} s_{x,i}(t - \tau_i)e^{j\phi_i} + N_0(t) \tag{3.1}$$

where $s_0(t)$ is the selected OFDM signal, $s_{x,i}(t)$ is the i-th interfering signal of N_x interferers and $N_0(t)$ is the ASE noise complex envelope. It is assumed here that, the ASE noise is following a zero mean Gaussian distribution with a variance of $N_0 B_{sim}$, where N_0 is the ASE noise power spectrum density and B_{sim} is the bandwidth used in the MC simulation, τ_i and ϕ_i are, respectively, the time delay and the phase difference between the selected and the i-th interfering signals. The τ_i parameter is considered as an uniformly distributed random variable between zero and T_s, and ϕ_i has a uniform distribution within the interval $[0, 2\pi]$ [27]. The crosstalk level is taken as the ratio between the average powers of the i-th interferer and the selected OFDM signal [28]. A sample function of ASE noise and in-band crosstalk are generated in each MC simulation and added to the optical OFDM signal.

When estimating the EVM, the MC simulation stops after 75 iterations [2], and then, the root mean square (rms) of the EVM, EVM_{rms}, of each k-th OFDM subcarrier is evaluated using [2]

$$EVM_{rms}[k] = \sqrt{\frac{\sum_{n=1}^{N_s} |s_r^n[k] - s_t^n[k]|^2}{\sum_{n=1}^{N_s} |s_t^n[k]|^2}} \quad k \in \{1, 2, ..., N_{sc}\} \tag{3.2}$$

where $s_r^n[k]$ and $s_t^n[k]$ are, respectively, the received and transmitted symbol at the k-th subcarrier of each n-th OFDM symbol of the total number of generated OFDM symbols, N_s. Then, the BER of each subcarrier, $BER[k]$ is calculated from [25]

$$BER[k] = 4 \frac{(1 - 1/\sqrt{M})}{\log_2(M)} Q \left(\sqrt{\frac{3}{(M-1) \cdot EVM_{rms}[k]^2}} \right) \qquad (3.3)$$

and the OFDM signal overall BER is given as follows

$$BER = \frac{1}{N_{sc}} \sum_{k=1}^{N_{sc}} BER[k] \qquad (3.4)$$

It should be remarked that (3.3) assumes a Gaussian distribution for the amplitude distortion at each OFDM subcarrier [2].

The BER is estimated from DEC after a total of 5000 counted errors, N_e, is reached in the OFDM received signal, and is obtained using $N_e/(N_s N_{it} N_{sc} N_b)$, where N_{it} is the number of iterations of the MC simulation and N_b is the number of bits per symbol in each OFDM subcarrier [2].

3.4   Results and Discussion

In this section, the tolerance to in-band crosstalk of the VC-assisted DD MB-OFDM communication system will be numerically assessed. The maximum tolerated crosstalk level, $X_{c,max}$, will be considered as the crosstalk level that leads to a 1 dB optical signal-to-noise ratio (OSNR) penalty. The OSNR penalty is defined as the difference in dB between the OSNR in presence of crosstalk and the OSNR without crosstalk that lead to a BER of 10^{-3} [28].

Table 3.1 displays the parameters used in MC simulation to assess the tolerance to in-band crosstalk of the VC-assisted DD OFDM system for 4, 16 and 64-QAM modulation formats in the OFDM subcarriers. The -3 dB bandwidth of the BS (2nd-order Super-Gaussian), and the modulation indexes are obtained from [4].

The evaluation of the tolerance to in-band crosstalk of the DD OFDM receiver starts with the estimation of the required OSNR to attain a 10^{-3} BER without in-band crosstalk. It was first assumed here that the selected and interferer OFDM signals have identical VBPR and VBG.

The plot of Fig. 3.5 shows the required OSNR as a function of the VBPR for a BER of 10^{-3} for different modulation format orders, without in-band crosstalk. It can be seen from Fig. 3.5, that the required OSNR for a BER of 10^{-3} increases almost linearly with the VBPR increase. This behavior is due to the SEB. As shown in Fig. 3.4, the SEB uses the OFDM subcarriers impaired by ASE noise to estimate and remove the SSBI from the OFDM signal. After the subtraction is performed, the ASE noise is partially removed and the OFDM performance degradation is solely determined by the VC-ASE beat noise in the received OFDM signal. Higher VBPR

Table 3.1 VC-assisted DD OFDM system simulation parameters

Modulation format	4-QAM	16-QAM	64-QAM
Bit rate per band [Gbps]	5.35	10.7	16.05
Number of subcarriers (N_{sc})	128		
Bandwidth per band [GHz]	2.7		
OFDM symbol duration [ns]	47.85		
Radio-frequency (f_{RF}) [GHz]	5		
Modulation index [%]	5		

Fig. 3.5 OSNR required values for a BER of 10^{-3} in the absence of in-band crosstalk as a function of VBPR, for different M-QAM modulation format orders

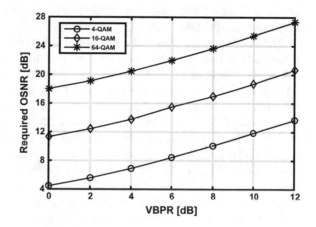

corresponds to a lower power on the OFDM band, thus, less tolerance to ASE noise. Therefore, in order to achieve the target BER, a higher OSNR is required.

The influence of the VBPR on the VC-assisted DD OFDM system performance is evaluated in Sect. 3.4.1. The impact of the VBG of the interferer on the in-band crosstalk tolerance is evaluated in Sect. 3.4.4, making use of two distinct scenarios as follows: scenario (a), the frequencies of the selected signal and interferer VCs are the same, and scenario (b) the central frequencies of the OFDM band of the selected signal and interferers have the same value.

3.4.1   Effect of the VBPR on the In-Band Crosstalk Tolerance

In this section, the effect of the VBPR of the interferer and selected signal on the tolerance to in-band crosstalk of DD OFDM receiver is addressed, taking into account two different situations. In Sect. 3.4.2, it is assumed that the selected and interfering signals have the 16-QAM modulation format in the OFDM subcarriers, but the VBPR of the interferer, $VBPR_x$, is changed. In Sect. 3.4.3, the study presented in Sect. 3.4.2 was extended by evaluating the VBPR impact on the tolerance to in-band crosstalk of the DD OFDM receiver for different modulation formats orders.

3.4.2 Same Modulation Format Order

In this section, the VBPR of the selected OFDM signal is set to 6 dB and the $VBPR_x$ is changed in the 0–12 dB range. The VBPR of 6 dB is achieved from the optimization carried out in [4].

The tolerated crosstalk level is assessed using the DEC and EVM methods. The accuracy of the EVM method is evaluated by comparing its estimates with the ones obtained using the DEC method.

The OSNR penalty is plotted in Fig. 3.6, obtained using the EVM method, versus crosstalk level due to a single interferer having different $VBPR_x$. Figure 3.6 shows that higher $VBPR_x$ lead to lower OSNR penalties. In Fig. 3.7, the tolerated crosstalk level of Fig. 3.6 is shown as solid line, as a function of the $VBPR_x$. In this plot the tolerated crosstalk level as a function of the $VBPR_x$, estimated using DEC, is also shown by the dashed line. A difference of about 2.8 dB between the tolerated crosstalk level for $VBPR_x$ of 0 and 12 dB can be observed, and the receiver performance degradation is seen to enhance with the $VBPR_x$ increase. In this case, the interfering band overlaps the selected OFDM band in the frequency domain, hence, the $VBPR_x$ increase reduces the power of its OFDM band, leading to less interference.

The tolerated crosstalk levels obtained using the DEC method are 0.8 to 1.8 dB higher than the ones obtained using the EVM method. Furthermore this difference revealed to be higher with the increase of the $VBPR_x$. For example, for 6 dB $VBPR_x$,

Fig. 3.6 OSNR penalty as a function of the crosstalk level obtained from the EVM method for 16-QAM modulation format, for interfering signals with different $VBPR_x$, and VBPR of the selected signal of 6 dB

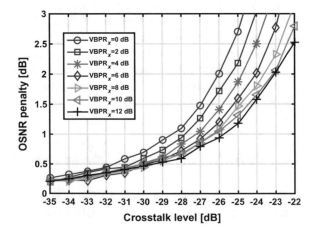

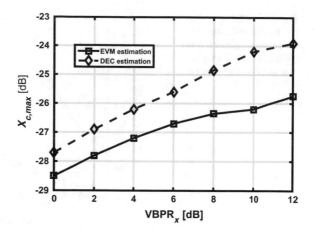

Fig. 3.7 Crosstalk tolerated level versus interferer VBPR, considering DEC (dashed lines) and EVM (solid lines) estimations

the EVM estimates a tolerated crosstalk level of -26.7 dB, while, the DEC predicts a tolerated crosstalk of -25.6 dB. Taking into account these differences between both methods, we can conclude that the EVM is imprecise on the estimation of the tolerated crosstalk level, when the selected and interferer OFDM signal have the same VBG but different VBPRs. As the in-band crosstalk sample functions are not modelled by a Gaussian distribution, the BER calculated from (3.3) can give rise to a imprecise BER predictions [2].

3.4.3 Different Modulation Format Orders

In this section, it is intended to address the influence of the VBPR on the tolerance to in-band crosstalk of the DD OFDM receiver for different modulation formats in the OFDM receivers. Since, in the previous section, it was concluded that the EVM method is inaccurate on the tolerated crosstalk level estimation for different VBPRs of the interfering and selected OFDM signals, the study presented in this section is based only on the DEC estimation of the DD OFDM receiver performance in presence of in-band crosstalk.

Figure 3.8 depicts the OSNR penalty as a function of the crosstalk level for 4, 16 and 64-QAM modulation format considering that the selected and the interfering OFDM signals have a VBPR of 6 dB. From Fig. 3.8, it can be concluded that the reduction of the modulation format order, used in the OFDM subcarriers, leads to an increase of the tolerance to in-band crosstalk of the DD OFDM receiver. This conclusion has been already reported [28], when the tolerance to in-band crosstalk of a M-QAM single carrier system with coherent detection was investigated. The 4-QAM OFDM signal is about 15 dB more tolerant to in-band crosstalk than the 64-QAM OFDM signal. Figure 3.8 shows that the tolerated crosstalk levels for the DD

Fig. 3.8 OSNR penalty
versus crosstalk level for
different modulation format
orders, for interferers and
selected signals having a
VBPR of 6 dB

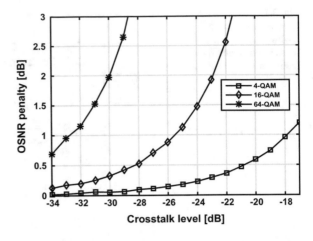

Fig. 3.9 Tolerated crosstalk
level versus VBPR$_x$ for
different modulation format
orders for 6 dB VBPR of the
selected OFDM signal

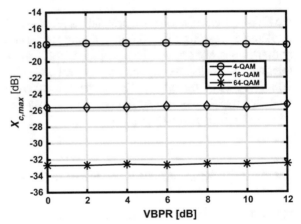

OFDM receiver are -18, -25.5 and -33 dB for the 4, 16 and 64-QAM modulation formats, respectively.

Figure 3.9 depicts the tolerated crosstalk level as a function of the VBPR for 4, 16 and 64-QAM modulation formats. Remark that, in this case, the VPBR of the interferer and the selected OFDM signals are the same. From Fig. 3.9, it can be observed that the tolerated crosstalk level is only dependent on the modulation format order of the OFDM subcarriers and it is essentially independent of the VBPR.

Figure 3.10 depicts the tolerated crosstalk level as a function of the VBPR$_x$, for different modulation format orders for a 6 dB VBPR level of the selected OFDM. In this figure it can be seen that the tolerated crosstalk level increases with the VBPR$_x$, as already seen in Sect. 3.4.2. For VBPR$_x$ of 0 dB, the VC-assisted DD OFDM system is 4 to 5 dB less tolerant to in-band crosstalk than the one estimated for VBPR$_x$ of 12 dB, regardless the format modulation order used in the OFDM subcarriers. Accordingly, by taking into account the conclusions arising from the results depicted

Fig. 3.10 Tolerated
crosstalk level versus
VBPR$_x$ for different
modulation format orders, a
6 db VBPR of the selected
OFDM signal is considered

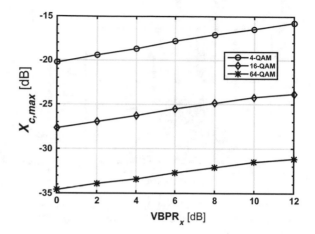

in Figs. 3.9 and 3.10, one can conclude that the tolerance to in-band crosstalk, for a
given modulation format order, is dependent on the difference between the VBPRs
of the selected and interferer OFDM signals. When the VBPR$_x$ is different from the
VBPR of the selected OFDM signal, the tolerance to in-band crosstalk is enhanced
with higher VBPR$_x$.

3.4.4   In-Band Crosstalk Tolerance Dependence on VBG

In this section, the influence of the VBG of the interferer on the VC-assisted DD
OFDM system performance is evaluated, at the light of two different scenarios. In
scenario (a), the frequency of the VC of the interfering OFDM signal is the same as
the VC of the selected signal, and the variation of the VBG of the interferer leads to
a frequency deviation between the central frequencies of the selected and interferer
OFDM bands, as it can be seen in graph of Fig. 3.11a. In the graph of Fig. 3.11a,
the spectrum of the selected OFDM signal is displayed in black while that of the
interfering OFDM signal spectrum is displayed in gray. The interferer has a VBG of
1.34 GHz and a crosstalk level of −20 dB. Figure 3.11b exemplifies the scenario (b),
in which, the central frequencies of the interfering and the selected OFDM bands
are the same, and the variation of the VBG of the interferer leads to a frequency
difference between the VC frequencies of the selected and interferer signals. In graph
of Fig. 3.11b, the VBG of the interferer is also 1.34 GHz. The influence of the VBG
on the tolerance to in-band crosstalk of the DD OFDM receiver is evaluated using
the DEC and EVM methodologies, and their estimations are confronted in order
to evaluate the influence of the VBG on EVM estimation accuracy of the tolerated
crosstalk level.

The plot of Fig. 3.12 represents the tolerated crosstalk level versus the interferer
VBG, for both simulation scenarios and estimated by the DEC method (dashed line)

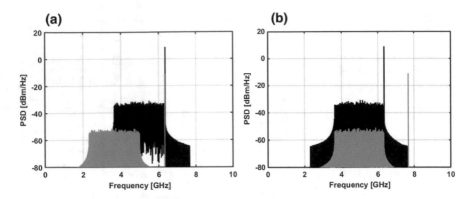

Fig. 3.11 PSDs of the selected OFDM signal (black) and interferer signal (gray) with a crosstalk level of −20 dB and a VBG = $B_w/2$ considering **a** same VCs frequencies and **b** equal OFDM bands central frequencies

Fig. 3.12 Tolerated crosstalk level versus VBG

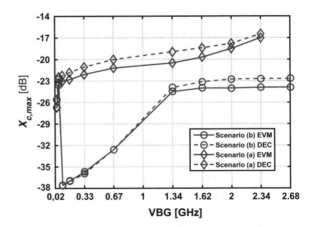

and EVM method (solid line). Focusing on scenario (a), Fig. 3.12 shows that the tolerance to in-band crosstalk increases with the interferer VBG. The crosstalk level with a VBG of 2.34 GHz is about 9 dB higher than the one obtained with a VBG of 20.9 MHz. As mentioned before, this scenario gives rise to the misalignment between the interferer central frequencies and selected bands. Therefore, as the VBG increases, less subcarriers of the selected band are being affected by the in-band crosstalk, and which causes the enhancement of the robustness of the DD OFDM system to in-band crosstalk. For a 2.68 GHz VBG, the interferer OFDM band is totally misaligned from the selected signal OFDM band, meaning that the OFDM subcarriers of the selected signal are not influenced by in-band crosstalk and leading to a null OSNR penalty. It can then be concluded that, in scenario (a), the DD OFDM receiver is completely tolerant to interfering OFDM signals with VBG equal or wider than the selected OFDM signal bandwidth.

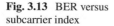

Fig. 3.13 BER versus
subcarrier index

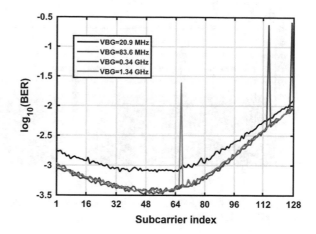

The graph of Fig. 3.12 also indicate that the EVM estimations for the tolerated crosstalk level are not in accordance with the DEC estimations, as a 2 dB difference between both estimations can be attained. Again, this disagreement between the two methods can be attributed to the non-Gaussian distribution of the in-band crosstalk sample functions.

Concerning scenario (b), Fig. 3.12 shows that the interfering signals having VBGs between 83.6 MHz and 0.67 GHz show a significant reduction on the tolerated crosstalk level, in comparison with smaller VBG. For a VBG of 83.6 MHz, the tolerated crosstalk level is about 11 dB smaller than the one obtained for a VBG of 20.9 MHz.

To go further in investigating this behavior, the BER estimated from the EVM is plotted versus the subcarrier index as shown in Fig. 3.13. In this figure, the crosstalk level was set to −26 dB, in order to allow a performance comparison with different interferer VBGs. Under this compliance it can be seen that for a 83.6 MHz VBG, the subcarrier 128 is the subcarrier with the worst performance due to the presence of the detected VC interferer. The increase of interferer VBG changes the subcarrier with less performance, in such a way that the BER per subcarrier is reduced, thus, decreasing the overall BER. The BS filtering was seen to account for this behavior. As the VBG increases, the frequency of the interferer VC becomes closer to the BS cut-off region, and thus, the VC power of is attenuated, so that after photodetection, the performance of the subcarrier that suffers the VC interference is incremented. For interferers having a VBG of 2.68 GHz, the frequency of the VC interferer is outside the BS passband, conducting to a full suppression of the interfering VC. This behavior can be outlined from the comparison of the OFDM signal spectrums at the photodetector output displayed in Fig. 3.14.

The graph of Fig. 3.14a displays the photodetection of the selected band in the presence of an interfering band having a VBG of 83.6 MHz, while, the one of Fig. 3.14b, displays the interferer band VBG is 1.34 GHz. By comparing the behavior of both graphs, it can be concluded that the VBG increase is giving rise to a frequency

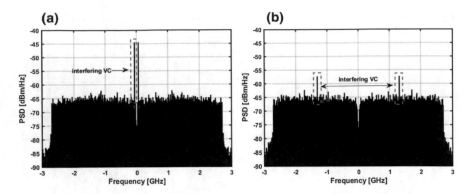

Fig. 3.14 PSDs of the photodetected signal for a crosstalk level of −20 dB for **a** VBG of 83.6 MHz and **b** VBG of 1.34 GHz

shift of the detected interferer VC and to an attenuation of its power imposed by the BS.

Finally, comparing the estimations of the tolerated crosstalk levels obtained from both methods, in scenario (b), Fig. 3.12 reveals that, for interferers with VBGs between 83.6 MHz and 0.67 GHz, the obtained tolerated crosstalk level estimations from the EVM and DEC are in compliance. Regarding remaining VBGs, for which higher tolerated crosstalk levels are estimated, a maximum difference of 1.2 dB between both estimations can be attained. It can then be asserted that the tolerated crosstalk level increase leads to a disagreement between the EVM and the DEC estimations, as the BER estimated from EVM, through (3.3), looses accuracy.

3.5 Conclusions

The OSNR penalty due to the in-band crosstalk on the performance of the VC-assisted OFDM metropolitan systems has been assessed using numerical simulations. The influences of the VBPR for different modulation format orders in the OFDM subcarriers and of the VBPR and the VBG of the interferer on the OSNR penalty have also been investigated.

It was shown that the influence of in-band crosstalk on the DD OFDM receiver performance depends on the difference between the VBPR of the selected and interferer OFDM signals and diminishes with the interferer VBPR increase. Higher VBPR was seen to lead to a reduction of the power of the interferer OFDM band, causing less interference on the detection of the selected signal. The increase of the modulation format order of the OFDM subcarriers also lead to less tolerance to in-band crosstalk. The 64-QAM modulation format has around 15 dB less tolerance to in-band crosstalk than the one found for the 4-QAM modulation format.

The impact of the interferer VBG on the in-band crosstalk tolerance, according with two different simulation scenarios, has also been analyzed. In a scenario (a), the VCs of the selected signal and interferer are set at the same frequency. In this case, results revealed that the increase of the overlapping between the interferer and selected OFDM subcarriers bands lead to higher system performance degradation. Larger VBG leads to a system performance improvement as the number of subcarriers that suffer in-band crosstalk is reduced. When the VBG of the interferer is equal to the selected OFDM signal bandwidth, the subcarriers of the interferer and selected OFDM signals are non-overlapped and the receiver performance is not degraded by in-band crosstalk. Regarding a scenario (b), selected and interferer OFDM bands are overlapped and the VBG of the interferer varied. For VBGs narrower than 1 GHz, the tolerance to in-band crosstalk is severely reduced and reaches a 11 dB less tolerance, for a VBG of 83.6 MHz. This behavior is caused by the detection of the interferer VC on the selected signal subcarriers that significantly degrades the DD OFDM receiver performance. For larger VBG, the VC of the interferer becomes closer to the cut-off region of the band selector, its power is attenuated and the performance of the DD OFDM receiver is slightly improved.

It has been also shown that the EVM method predicts inaccurate estimates of the maximum tolerated crosstalk level.

Acknowledgements This work was supported by Fundação para a Ciência e Tecnologia (FCT) from Portugal through Project MORFEUS-PTDC/EEITEL/2573/2012. The project UID/EEA/50008/2013 is also acknowledged.

References

1. T. Alves, A. Cartaxo, Performance degradation due to OFDM-UWB radio signal transmission along dispersive single-mode fiber. IEEE Photonics Technol. Lett. **21**(3), 158–160 (2009)
2. T. Alves, A. Cartaxo, Analysis of methods of performance evaluation of direct-detection OFDM communication systems. Fiber Integr. Opt. **29**(3), 170–186 (2010)
3. T. Alves, A. Cartaxo, Direct-detection multi-band OFDM metro networks employing virtual carriers and low receiver bandwidth, in *Optical Fiber Communication Conference*, San Francisco, USA (2014)
4. T. Alves, A. Cartaxo, High granularity multiband OFDM virtual carrier-assisted direct-detection metro networks. J. Lightwave Technol. **33**(1), 42–54 (2015)
5. J. Attard, J. Mitchell, C. Rasmussen, Performance analysis of interferometric noise due to unequally powered interferers in optical networks. J. Lightwave Technol. **23**(15), 1692–1703 (2005)
6. S. Blouza, J. Karaki, N. Brochier, E. Rouzic, E. Pincemin, B. Cousin, Multi-band OFDM for optical networking, in *EUROCON 2011—International Conference on Computer as a Tool—Joint with Conftele 2011*, Lisbon, Portugal (2011)
7. L. Cancela, J. Pires, Quantifying the influence of crosstalk-crosstalk beat noise in optical DPSK systems, in *International Conference on Computer as a Tool (EUROCON)*, Lisbon, Portugal (2011)
8. L. Cancela, J. Rebola, J. Pires, Implications of in-band crosstalk on DQPSK signals in ROADM-based metropolitan optical networks. Opt. Switch. Netw. **19**(3) (2016)

9. N. Cvijetic, OFDM for next-generation optical access networks. J. Lightwave Technol. **30**(4), 384–398 (2012)
10. R. Davey, D. Grossman, M. Rasztovits-Wiech, D. Payne, D. Nesset, A.E. Kelly, A. Rafel, S. Appathurai, S.i Yang, Long-reach passive optical networks. J. Lightwave Technol. **27**(3), 273–291 (2009)
11. M. Deng, Y. Xingwen, Z. Hongbo, L. Zhenyu, Q. Kun, Nonlinearity impact on in-band crosstalk penalties for 112-gb/s PDM-COOFDM transmission, in *Communications and Photonics Conference (ACP)*, Guangzhou (Canton), China (2012), pp. 1–3
12. M. Filer, S. Tibuleac, Generalized weighted crosstalk for DWDM systems with cascaded wavelength-selective switches. Opt. Express **20**(16), 17620–31 (2012)
13. K. Ho, Effects of homodyne crosstalk on dual-polarization QPSK signals. J. Lightwave Technol. **29**(1), 124–131 (2011)
14. M. Jeruchim, P. Balaban, K. Shanmugan (eds.), *Simulation of Communication Systems: Modeling, Methodology and Techniques*, 2nd edn. (Kluwer Academic Publishers, Norwell, MA, USA, 2000)
15. P. Legg, M. Tur, I. Andonovic, Solution paths to limit interferometric noise induced performance degradation in ASK/direct detection lightwave networks. J. Lightwave Technol. **14**(9), 1943–1954 (1996)
16. I. Monroy, E. Tangdiongga, *Crosstalk in WDM Communication Networks*, 1st edn. (Springer, New York, 2002)
17. S. Nezamalhossein, L. Chen, Q. Zhuge, M. Malekiha, F. Marvasti, D. Plant, Theoretical and experimental investigation of direct detection optical OFDM transmission using beat interference cancellation receiver. Opt. Express **21**(13), 15237–15246 (2013)
18. J. Pan, P. Isautier, M. Filer, S. Tibuleac, E. Ralph, Gaussian noise model aided in-band crosstalk analysis in ROADM-enabled DWDM networks, in *Conference on Optical Fiber Communication, Technical Digest Series*, San Francisco, USA (2014)
19. W. Peng, B. Zhang, K.M. Feng, X. Wu, A.E. Willner, S. Chi, Spectrally efficient direct-detected OFDM transmission incorporating a tunable frequency gap and an iterative detection techniques. J. Lightwave Technol. **27**(24), 5723–5735 (2009)
20. B. Pinheiro, J. Rebola, L. Cancela, Impact of in-band crosstalk signals with different duty-cycles in M-QAM optical coherent receivers, in *20th European Conference on Networks and Optical Communications (NOC)*, London, United Kingdom (2015), pp. 1–6
21. B. Pinheiro, J. Rebola, L. Cancela, Assessing the impact of in-band crosstalk on the performance of M-QAM coherent receivers using the error vector magnitude, in *10th Conference on Telecommunications (Conftele)*, Aveiro, Portugal (2015)
22. J. Rebola, A. Cartaxo, Impact of in-band crosstalk due to mixed modulation formats with multiple line rates on direct-detection OFDM optical networks performance, in *16th International Conference on Transparent Optical Networks (ICTON)*, Graz, Austria (2014)
23. J. Rebola, A. Cartaxo, Estimating the performance of DD OFDM optical systems impaired by in-band crosstalk using the moment generating function. J. Lightwave Technol. **34**(10), 2562–2570 (2016)
24. R. Schmogrow, Error vector magnitude as a performance measure for advanced modulation formats. IEEE Photonics Technol. Lett. **24**(1), 61–63 (2012)
25. R. Shafik, S. Rahman, R. Islam, On the extended relationships among EVM, BER and SNR as performance metrics, in *International Conference on Electrical and ComputerEngineering (ICECE)*, Dhaka, Bangladesh (2006), pp. 408–411
26. W. Shieh, I. Djordjevic, *OFDM for Optical Communications*, 1st edn. (Academic Press Elsevier, Burlington, MA, 2010)
27. P. Winzer, M. Pfennigbauer, R.-J. Essiambre, Coherent crosstalk in ultradense WDM systems. J. Lightwave Technol. **23**(4), 1734–1744 (2005)
28. P. Winzer, A.H. Gnauck, A. Konczykowska, F. Jorge, J.-Y. Dupuy, Penalties from in-band crosstalk for advanced optical modulation formats, in *37th European Conference and Exhibition on Optical Communication*, (1), Geneva, Switzerland (2011), pp. 1–3
29. T. Xia, G. Wellbrock, *Commercial 100-Gbit/s Coherent Transmission Systems, Optical Fiber Telecommunications*, 1st edn. (Academic Press, Oxford, 2013)

Chapter 4
High Accuracy Laser Power Measurement for Laser Pointers Safety Assessment

Kanokwan Nontapot and Narat Rujirat

Abstract Laser pointers are common tools for teachers, students, and researchers as pointer devices in classrooms or meeting rooms. Some people use laser pointers for toys, hobbies and entertainments. For safety concern, these laser pointers are designed to emit laser power level below 5 mW as restricted by many countries regulations. However, advances in laser technology, allowed the production of low low-cost laser pointer devices, at a wide range of visible wavelengths, delivering considerable light output power, which are available for the general public. In fact it is now common, to have laser pointers and laser gadgets of all colors and power range, from few mW to several watts available to buy. As a result, laser pointers are in the hands of uninformed people about the potential injuries that can arise from the handling of these devices. This is resulting in a worldwide increase of retinal injuries reporting. In this research, the laser power levels of randomly purchased laser pointers in Thailand markets will be assessed in order to check the accuracy of their laser safety labels, using a dedicated laser pointer power testing kit developed at National Institute of Metrology (Thailand). A set of twenty laser pointers, red, green and violet, randomly acquired in market have been assessed with respect to their power output level and emitted wavelengths lines. The limits imposed by The US Code of Federal regulations for lasers devices will be used as a reference.

4.1 Introduction

Laser pointing devices are being widely used over the world in a wide range of applications. The most often seen are the laser pointers devices used in presentations and demonstrations, particularly in classroom and conferences. They are also being used in entertainment as toys and special light effects in several public events. These devices become accessible to public in a wide range of colors and power

K. Nontapot (✉) · N. Rujirat
National Institute of Metrology, Pathum Thani, Thailand
e-mail: kanokwan@nimt.or.th

N. Rujirat
e-mail: narat@nimt.or.th

© Springer Nature Switzerland AG 2018
P. A. Ribeiro and M. Raposo (eds.), *Optics, Photonics and Laser Technology*, Springer Series in Optical Sciences 218, https://doi.org/10.1007/978-3-319-98548-0_4

level at very accessible prices. Nowadays red, green and violet laser pointers are the most vulgarized within the public and more recently the blue ones are taking place. Concerning to laser light power coming out of such a devices dangerous levels as high as 1 W can be freely acquired in market although there are limits imposed in majority of countries. For example 5 mW is the limit in the States and 1 mW in UK. The practically free access to the public of high power laser pointer devices is not apart from the increase in eyes injury reporting over the world [1–3]. Laser pointers, due to their long range, can be considered extremely hazardous if pointed at an aircraft. In 2010, there were over 2,836 cases of aviation-related laser pointer incidents report by U.S. Federal Aviation Administration (FAA) according to MSNBC report (msnbc.msn.com, 2011). For example in July 1 2016, there have been over 42,000 cases of laser illuminations reported by pilots to the U.S. FAA and the U.K. Civil Aviation Authority, since 2004. So far, the most recent and most severe impact of laser pointer to aviation is a flight of British Airways which was aborted after an hour of take-off in February 14 2016 (http://www.bbc.com/news/uk-35575861). Laser illumination incidents reported by pilots to the U.S. Federal Aviation Administration from 2004 to 2016, which are increased significantly over the past few years, are shown in Fig. 4.1 (www.laserpointysafety.com). There were also many cases of laser pointers that were aimed at players' eyes in worldwide sport events especially in soccer games (www.laserpointysafety.com). Under this compliance governments started to rule laser marked. For example, in 2008 the Australian government limited both sale and importation of some laser products and limited allowed output power of imported laser pointer devices to 1 mW. This restriction was mainly motivated by the registered cases of coordinated attacks to passenger jets in Sydney [4].

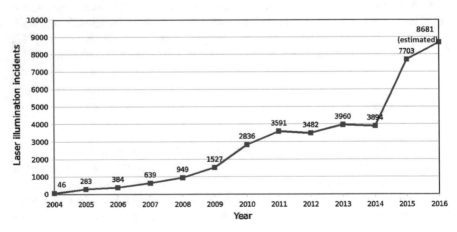

Fig. 4.1 Laser illumination incidents reported by pilots to the U.S. Federal Aviation Administration

4.1.1 Related Studies of Laser Pointer Safety Assessments

In 2010, the Federal Office of Metrology in Switzerland randomly tested laser pointers and found that high percentage of laser pointers were not in compliance with the regulation [5]. In 2011, the U.K. Health Protection Agency (HPA) tested 246 laser pointers seized from a company suspected of violations import [6]. The tested from HPA found that 96% of the working pointers were above the U.K.'s legal power limit of 1 mW. As a result, 7,378 of laser pointers were seized, together with 8,780 assembled parts. It was estimated that the company sold over 35,000 laser pointers from 2009 to 2011 and generating income over USD 1,600,000. Out of 246 pointers tested by HPA, 9 were under 1 mW limit and compliance to the U.K. legal limit, 27 were not working, 14 emitted power between 1 and 5 mW and 196 laser pointers emitted power above 5 mW. The study showed that approximately 80% of the lasers tested were potentially hazard to users.

In 2013, Trevor Wheatley, chair of the Standards Australia SF-019 Committee on laser safety studied 41 lasers pointers purchased online which were claimed by the sellers to be legal and emit laser power lower than the Australian limit of 1 mW. The study found that 95% of these pointers emit power more than 1 mW, which were illegal under Australian law and 78% were emit laser power between 5 mW and 100 mW [7].

Researchers at National Institute of Standard and Technology (NIST), USA, tested 23 laser pointers and found that 44% of red pointers and 90% of green laser pointers were not compliance with federal safety regulations [8]. And most recently, more comprehensive test had been conducted on 122 laser pointers, most purchased online. The test results reveal that 44% of red laser pointers and 90% of the green laser pointers did not comply with U.S. safety regulations. Moreover, green laser pointers also emitted dangerous invisible infrared radiation, in addition to the visible green light [9]

The website www.laserpointersafety.com reported that in December 2012, 24 laser pointers being sold on Amazon.com were tested by a laser manufacturer. The results revealed that all laser pointers were over legal limit and thus illegal. Those laser pointers were advertised as being 5 mW or less. However, the actual average emitted power was 41 mW. All of the green laser pointers under tested also emitted dangerous invisible infrared radiation, in addition to the visible green wavelength. The results of the tested are shown in Table 4.1.

In some countries as Thailand for example, trading of laser devices is not yet regulated. Thus there are no restrictions with respect to either device optical power or emission wavelengths, in such a way that laser pointers and laser gadgets can be found in market in practically all available wavelengths and power ranges, from less than mW to many watts. Thus, improvident or even the malicious uses of these devices can have very serious consequences. For all of these reasons it is fundamental to assess the power output of laser devices available in market in order to determine their real hazard level and provide the correct device class labels. In this work, 19 laser pointers of 3 different colors, namely red, green and violet, randomly acquired on stores and online, were assessed with respect to their power levels and wavelengths of

Table 4.1 Test results of laser power from laser pointers sold on Amazon.com

Wavelength (nm)	Claim output power (mW)	Actual output power (mW)	Infrared output power (mW)
650	≤5	111.9	–
405	≤5	87.6	–
532	≤5	80.1	19.8
532	≤5	74.4	29.4
405	≤5	69.6	–
532	≤5	49	40.5

light emitted in order to check their hazard level. The US Code of Federal Regulations [10] was used as standard for hazard level of laser pointers which limits commercial class IIIa/3R lasers to 5 mW and no more than 1 mW for dangerous IR.

4.1.2   Laser Class and Safety

Laser Safety. Laser safety concerns with the secure use and operation of lasers towards risk minimization as a result of laser incidence, particularly with respect to eyes injuries. The sale and usage of laser is usually subject to governments regulations since only a small amount of laser can lead to potential hazard to eyes. Medium and high-power laser can be extremely hazardous since they can burn eyes retina, or even skin.

Maximum Permissible Exposure. The maximum permissible exposure figure of merit (MPE) is the maximum light source power density (W/cm^2) or energy density (J/cm^2) levels established as safe to human eye or skin at a given wavelength and exposure time. MPE is calculated from worst-case scenario where the light of each wavelength is focused into the smallest spot size on the retina under fully open pupil conditions (0.39 cm^2) for visible and near-infrared wavelengths. Methods for MPEs calculation can be found in the IEC-60825-1 and ANSI Z136.1 standards [11, 12]. MPE expressed as energy density versus exposure time and MPE or as power density versus exposure time for different wavelengths, have been plotted by Han-Kwang [13], in following IEC 60825 standard. These plots are shown in Figs. 4.2 and 4.3. The graph show that deep UV light, at very low power, can be very dangerous if stare for long time. For laser light in infrared region, MPE is higher than that of visible light because light at these wavelengths can be absorbed by the transparent parts of the eye, before they reach the retina.

4.1.3   Laser Class

The American Standard for laser use safety [12] divides lasers into four classes, according with their hazard level as shown in Table 4.2.

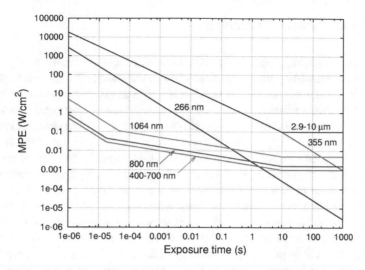

Fig. 4.2 MPE as power density versus exposure time for different wavelengths

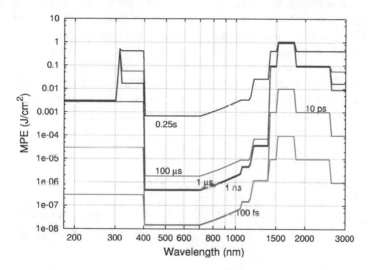

Fig. 4.3 MPE as energy density versus exposure time for different wavelengths

Table 4.2 Laser classification according with ANSI Z136.1-2007

Classes	Hazard Level
Class 1	Non-hazardous level under normal use
Class 2	Non-hazardous level under normal use, for visible laser only, laser power <1 mW
Class 3R	Potentially hazardous level if viewing for extend period of time, includes both visible and non-visible lasers, laser power <5 mW
Class 3B	Hazardous to skin and eyes, includes both visible and non-visible lasers, laser power <500 mW
Class 4	Very hazardous, laser power >500 mW, can burn skin

4.2   Laser Pointer Technology

Actual laser technology allowed the achievement of a wide variety of high performance laser assemblies which are accessible to the public. For example, red laser pointers in present days, are built based from vertical-cavity surface-emitting laser (VCSEL) diodes, emitting light near 650 nm, are more efficient and deliver more power than the early generation 633 nm He-Ne gas lasers. In turn, green laser pointers, most popular as human eyes are most sensitive in the green region of the spectrum, are in fact for diode pumped solid state (DPSS) lasers, or diode pumped solid state frequency-doubled (DPSSFD). In this case the green laser light is indirectly generated by means of an infrared AlGaAs laser diode operating at 808 nm pumping a Nd:YAG or a Nd:YVO4 crystal. The laser emission takes place at 1064 nm is then converted into 532 nm green light by using second harmonic generation KTP nonlinear optical crystal. Actually three different wavelengths, IR (808 and 1064 nm) and VIS (532 nm), will in principle be coming out from the device, unless proper filtering is implemented. Thus, safe green laser device pointers must incorporate optical filters to attenuate both IR wavelengths present. According with ANSI [12] and IEC [14] standards power levels at output should be less than 0.63 mW at 808 nm and less than 1.92 mW at 1064 nm respectively. With respect to violet laser pointers, usually emitting at 405 nm, they are built from GaN (gallium nitride) semiconductors. This type of laser devices directly emit 405 nm light without requiring frequency doubling crystal as in the DPSS green laser devices. Thus in this case none of the dangerous IR light is emitted however output power in excess of the CFR limit could be harmful at this near UV wavelength. Table 4.3 displays some common laser pointers and their wavelengths.

Table 4.3 Common laser pointers and their wavelengths

Wavelength (nm)	Laser types
405	InGaN Blue-violet laser, in blue ray disc and HD DVD drive
445	InGaN blue laser multimode diode for use in data projectors
510–525	Green laser diodes developed by Nichia and OSRAM for laser projectors
635	AlGaInP, brighter red laser pointers, with same power but twice as bright as 650 nm
650–660	GaInP/AlGaInP in CD DVD, cheap red laser pointers
670	AlGaInP bar code readers, first diode laser pointers (now obsolete)
808	GaAlAs pumps in DPSS Nd:YAG lasers (in green laser pointers)
848	Laser mice
980	InGaAs pump for optical amplifiers, for Yb:YAG DPSS lasers
1,064	AlGaAs fiber-optic communication, DPSS laser pump frequency

4.3   Experimental Assembly

It is therefore essential to address the development of dedicated optical apparatus, accurate enough to measure the output power level of laser pointer devices at the different wavelengths emitted. Thus the apparatus should be able to measure not only the light power emitted in the laser pointer visible fundamental wavelengths, red, green and violet, but also in IR regions. The design has to be able for use with different type of laser pointers.

4.3.1   Optical Bench

The apparatus has to be built according with safety requirements of operation and should be able of accurate and repeatable results. The optical layout, outlined in Fig. 4.4, was inspired in Joshuas prototype [8]. It consists of a power meter, a set of selectable filters, self-centering cage type mounts and an adjustable iris, Fig. 4.4.

4.3.2   Laser Power Detector

Up to date, there are 3 main types of laser power detectors, which are thermopiles, pyroelectric sensors, and photodiodes.

Thermopile Sensors. Thermopile sensors are used for measuring CW laser power and integrates energy of pulsed laser to produce an average laser power measurement. Thermopiles sensors absorb radiation from incident laser beam and convert into heat.

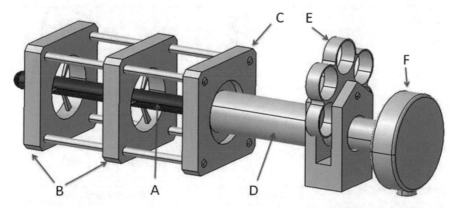

Fig. 4.4 Optical bench for laser pointer assessment: (A) laser pointer; (B) laser pointer mount; (C) adjustable iris; (D) lens tube; (E) Bandpass filters set mounted in a selectable filter wheel; (F) power meter

The heat is then flows to a heat sink that is kept at ambient temperature by mean of water-cooling or air-cooling. A thermopile junction then transduces the temperature difference between the heat sink and absorber into electrical signal. Since thermopile converts thermal energy into electrical energy, it offers spectrally flat responsivity.

Pyroelectric Sensors. Pyroelectric sensors are used to measure the energy of pulsed lasers and can only be used with pulsed laser. They can also be used to measuring average power of pulsed laser by multiplying laser pulse energy by the laser repetition rate. Pyroelectric sensors are thermal sensors in which fluctuations in temperature caused by absorption of laser light produce a charge change on the surface of pyroelectric crystals, which then convert into an electrical signal.

Semiconductor Photodiodes. Semiconductor photodiode sensors directly convert incident photon into electrical current. They are used for measuring low CW laser power. Photodiode offers fast response time, however, the spectral range is more limited than for other. Example of spectral responsivity of a silicon photo diode is shown in Fig. 4.5 [15]. Commonly used photodiodes are from silicon, germanium and indium gallium arsenide semiconductors. Photodiode materials and their spectral ranges are shown in Table 4.4.

A thermopile detector has been used to measure laser pointers light power, mainly due to its spectral flat responsivity, easy to implement and low cost. The detector thermopile, Ophir Model 3A-P covers the spectral and power ranges of 0.15 μm–6 μm, and 60 μW–3 W respectively. Detector calibration factor and determination of respective uncertainties, were carried out using standard calibration procedure of laser power detectors at Thailand's National Institute of Metrology by comparison with a laser power reference standard unit. The details of calibration procedure are presented in Appendix.

Fig. 4.5 Spectral responsivity of a silicon photo diode

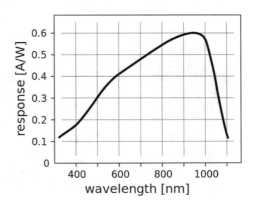

Table 4.4 Photodiode materials and their spectral ranges

Material	Spectral range (nm)
Silicon	190–1100
Germanium	400–1700
Indium gallium arsenide	800–2600

4.3.3   Bandpass Filter

As mentioned above the DPSS green laser pointers emit laser light not only at visible green (532 nm) but also in the IR region at the pump wavelength 808 nm and fundamental 1064 nm. To measure the emitted power at each of these wavelengths a set of bandpass filters was used. The filters central lines and Full Width at Half Maximum (FWHM) of filters used for measuring the 800 nm and 1064 nm wavelengths were 800 ± 8 nm, FWHM = 40 ± 8 nm and 1064 ± 2 nm, FWHM = 10 ± 2 nm respectively. The 1064 nm filter transmission was checked with a Nd:YAG laser beam. The 800 nm filter was also checked and assumed to have the same uncertainty. The obtained transmission values for the 800 nm filter in the 600 nm to 1000 nm range and the ones for the 1064 nm filter in the 1000 nm to 1150 nm range are plotted Figs. 4.6 and 4.7 respectively.

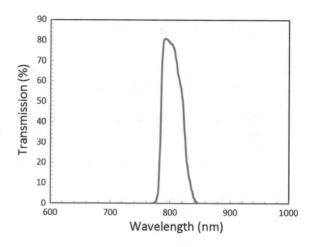

Fig. 4.6 Measured transmission values for 800 nm filter from 600 to 1000 nm

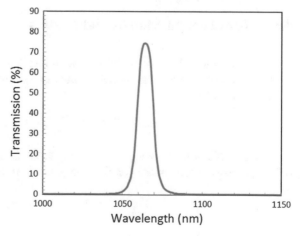

Fig. 4.7 Measured transmission values for 1064 nm filter from 1000 to 1150 nm

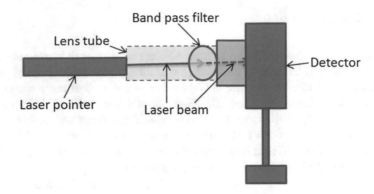

Fig. 4.8 Diagram of laser pointer measurement and laser beam paths. Solid line represent laser beam before travelling to bandpass filter. Dash line represent transmitted laser beam from bandpass filter

4.4 Measurement Procedure

Diagram for laser pointer power measurement and laser beam path are shown in Fig. 4.8. The procedure for measuring laser power emit from laser pointers are as follows:

1. Mount the laser pointer using self-centering lens mounts (Fig. 4.4).
2. Select a bandpass filter to be used by rotating a selectable filter wheel to a desired filter.
3. Energize laser pointer for 30 s and record the maximum power from the laser meter.
4. Repeat the measurement 5 times for each laser wavelength. Calculate the average power and standard deviation.

4.5 Measurement Models and Results

For single wavelength lasers as red and violet laser pointers, no bandpass filters are required for laser power measurement. Under this situation the device output power is simply given by

$$P_{red,violet} = \frac{P_{total}}{C_n} \tag{4.1}$$

For the case of multi-wavelength emitting devices, as green laser pointers, the total output power will contain the contribution of all wavelengths, and can be expressed as:

$$P_{532} = \frac{P_{total}}{C_n} - [P_{808} + P_{1064}] \tag{4.2}$$

$$P_{808} = \frac{P'_{808}}{C_n T_{808}} \tag{4.3}$$

$$P_{1064} = \frac{P'_{1064}}{C_n T_{1064}} \tag{4.4}$$

where C_n is a calibration term for the power meter, according with calibration procedure (see appendix). P'_λ is the power of the laser wavelength λ from the detector and T_λ is the transmission of bandpass filter at wavelength λ. For each laser pointer device the emitted power output was recorded for 30 s power using the maximum power reading feature of the power meter. A total of 19 laser pointer devices (12 red, 4 green, and 3 violet), randomly acquired Thailand's stores and online market have been analyzed in this way. All of the pointers were pointed out as demonstration purposes laser pointer devices and all claimed to be in class 3R or below. Measurement outcomes are shown in bar graphs of Figs. 4.9, 4.10 and 4.11. For the red laser pointer devices, 3 of the 12 emitted laser power light above 5 mW, thus exceeding the CFR limits. Results for the green laser pointers, revealed that 2 out of 4 emitted power above CFR limits - Class 3R visible accessible emission limit (AEL) and Class 1IR AEL. With respect to the visible green light, one of the devices revealed emitted ∼6 mW light power and another reached a power level as high as ∼70 mW. At 1064 nm fundamental wavelength, one of the pointer devices presented a ∼10 mW output

Fig. 4.9 Output power level outcomes of 12 red laser pointer devices. The horizontal dashed line is indicating the 5 mW limit corresponding to the class 3R visible AEL. Results are showing that 3 out of 12 red pointers emitted power in excess of the CFR limits

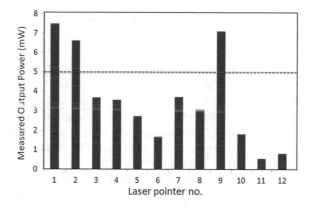

Fig. 4.10 Outcomes for the output power of 4 green laser pointer devices. The horizontal dashed line shows the class 3R visible AEL limit of 5 mW. As it can be seen 2 of the 4 green pointers emitted power exceeding the CFR limits with respect to both visible and IR wavelengths

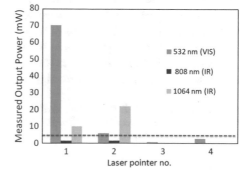

Fig. 4.11 Measured output power of 3 violet laser pointers. The horizontal dashed line represents the class 3R visible AEL limit of 5 mW. All of the violet pointers emitted power in excess of the CFR limits

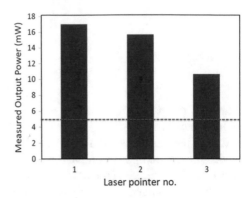

and another reached the level of ∼20 mW. With respect to emission at the IR 808 nm pump wavelength, both laser devices delivered ∼1.5 mW output power level. Finally with respect to the three violet laser pointers assessed, all advertised to be below 5 mW output level, outcomes revealed that all emitted the power in excess to the CFR limit.

4.6 Conclusions

This study is providing the first and most accurate results with respect to the laser pointers safety assessment level in Thailand, a country in which there is no regulation regarding the trading of laser devices. The power outputs of 19 laser pointer devices of the 3 main available colors, red, green and violet, randomly acquired from market, were assessed to check the conformity with US CFR and ANSI standards. The uncertainty of power measurements was determined from the calibration of both power detector and the bandpass filter. The calibration of power detector was carried out according with standard practice of the Thai National Institute of Metrology laser power calibration, traceable to SI unit. Calibration of bandpass filters were in compliance with the procedure outlined by Joshua [8]. The overall attained uncertainty with respect to power measurement of used experimental setup and procedure was of 1%. Results of the laser pointer devices assessment power showed that the majority of red laser devices are in compliance with the CFR regulations, while 50% of green laser pointers exceeded 3R class limit at both visible and IR wavelengths. For the violet lasers pointers, all of them were non-compliant with CFR limits. Finally, it was shown that the apparatus here described and results obtained from it could be used as guidance for regulation and standardization bodies, when issuing regulations regarding the sale or importation of laser devices.

Acknowledgements This work is supported by the new researcher scholarship of CSTS, MOST project, which is sponsored by the Coordinating Center for Thai Government Science and Technology Scholarship Students (CSTS), National Science and Technology Development Agency (NSTDA).

Appendix

Measurement Results

Average laser output power outcomes from red, green, and violet laser pointer devices are shown in Tables 4.5, 4.6 and 4.7.

Measurement Uncertainty. Measurement uncertainties were estimated following the guideline from Taylor and Kuyatt [16] and Evaluation of measurement data, Guide to the expression of uncertainty in measurement (GUM) [17]. There are 2 types of uncertainties associated with measurement, type A standard uncertainty, arising from statistical analysis of a series of observations, and type B standard uncertainty, coming from other than statistical analysis, for this work, in this case related with the measuring system itself.

Type A Uncertainty. Calculated from standard deviation of 5 laser power measurements, S_r. Type A standard uncertainty corresponds to standard deviation of the mean $\frac{S_r}{\sqrt{5}}$.

Type B Uncertainty. Obtained from uncertainty of measurement system. In this work, the estimated system uncertainty, u_s, is

$$u_s = \sqrt{u_d^2 + u_f^2} \tag{4.5}$$

where u_d is the power detector uncertainty of and u_f is the bandpass filter uncertainty. The combined standard uncertainty of measurement, U, is

$$U = \sqrt{u_s^2 + \frac{s_r^2}{5}} \tag{4.6}$$

Table 4.5 Average laser output power from red laser pointers

Laser pointer no.	Average output power (mW)
1	7.48
2	6.60
3	3.66
4	3.54
5	2.71
6	1.64
7	3.70
8	3.10
9	7.10
10	1.80
11	0.55
12	0.80

Table 4.6 Average laser output power from green laser pointers

Laser pointer no.	Average power without filter (mW)	Average power of visible wavelength 532 nm (mW)	Average power of IR wavelength 808 nm (mW)	Average power of IR wavelength 1064 nm (mW)
1	81.60	69.94	1.51	10.15
2	29.50	5.92	1.48	22.1
3	0.70	0.70	0	0
4	2.8	2.8	0	0

Table 4.7 Average laser output power from violet laser pointers

Laser pointer no.	Average output power (mW)
1	17
2	15.7
3	10.7

The uncertainty of the power detector is obtained by calibration of laser reference standard, by direct comparison with the laser Calorimeter (traceable to SI unit). Calibration was carried out at 633, 515 and 488 nm wavelengths. The calibration experimental setup is shown in Fig. 4.12. The calibration factor (C_n) is

$$C_n = \frac{P_u}{P_s} \tag{4.7}$$

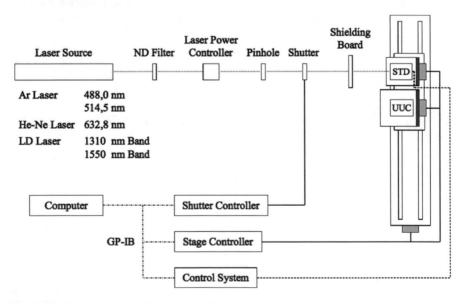

Fig. 4.12 The laser power calibration experimental setup

Table 4.8 Power detector calibration results

Wavelength (nm)	Power (mW)	Calibration Factor, C_n	Uncertainty, u_d (%)
633	1	1.015	0.6
	3	1.012	0.6
515	5	1.025	0.6
	10	1.025	0.6
488	5	1.009	0.6

Fig. 4.13 The bandpass filter calibration experimental setup

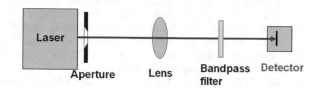

Table 4.9 Bandpass filter calibration result

Filter center wavelength (nm)	Laser wavelength (nm)	% Transmission	Uncertainty, u_f
800	532	0.00	–
	1064	0.00	–
1064	532	0.00	–
	1064	84.92%	0.1%

where P_u is power measured by the unit under test (thermopile detector) [W], and P_s is the laser power measure by the standard [W]. The calibration results are shown in Table 4.8.

The uncertainties of the bandpass filters were estimated from transmission calibration measurements using a Nd:YAG laser light source at 1064 nm and 532 nm. The calibration experimental setup is shown in Fig. 4.13. For the 1064 nm bandpass filter, the transmission value considered was the ratio of laser power reading on a standard pyroelectric detector with and without the filter. The calibration attained values are displayed shown in Table 4.3. Concerning the the 800 nm filter, the light transmissions at 1064 nm and 532 nm were checked. The calibration results are in Table 4.9. By using (4.6), a 1% of total uncertainty can be obtained.

References

1. R. Wong, D. Sim, R. Rajendram, G. Menon, Class 3A laser pointer-induced retinal damage captured on optical coherence tomography. Acta Ophthalmol. Scand. **85**(2), 227–228 (2006)
2. S. Wyrsch, P.B. Baenninger, M.K. Schmid, Retinal injuries from a handheld laser pointer. N. Engl. J. Med. **363**(11), 1089–1091 (2010)

3. K. Ziahosseini, J.P. Doris, G.S. Turner, Maculopathy from handheld green diode laser pointer. BMJ **340**(1), c2982–c2982 (2010)
4. Sydney Morning Herald, Laser Pointers Restricted After Attacks (2008)
5. P. Blattner, The underrated hazard potential of laser pointers, in *METinfo18* (2011)
6. H. John, H. Michael, K. Marina, Laser product assessment for Lancashire County Council Trading Standards Service, in *Proceedings of the 2013 International Laser Safety* (2013)
7. W. Trevor, *Proceedings of the 2013 International Laser Safety Conference*, Orlando, FL (2013), pp. 48–54
8. H. Joshua, D. Marla, Accurate, inexpensive testing of laser pointer for safe operation. Meas. Sci. Technol. **24**, 045202 (7 pp.) (2013)
9. H. Joshua, Random testing reveals excessive power in commercial laser pointers. Presentation at the International Laser Safety Conference, Orlando, FL, 20 Mar 2013
10. Performance Standards for Light-Emitting Products, *CFR 2012 Code of Federal Regualtions*, Title 21. Food and Drug; Part 1040; Section 1040.10 Laser Products (US 21 CFR 1040.10)
11. International Electrotechnical Commission, *IEC 2007 Safety of Laser Products. Part 1: Equipment classification and requirements*, EN 60825-1 ed.2 (2007)
12. ANSI 2007 Laser Institute of America, *American National Standard for the Safe Use of Lasers ANSI Z136.1*, Orlando (2007)
13. K. Han, IEC60825 MPE Js, image (2007), https://commons.wikimedia.org/w/index.php?curid=17862637
14. International Electrotechnical Commission, *Safety of laser products—Part 1: Equipment classification and requirements* (2007)
15. M. Kai, Response silicon photodiode, image (2009), https://commons.wikimedia.org/w/index.php?curid=6625679
16. N. Taylor, C. Kuyatt, Guidelines for Evaluating and Expressing the Uncertainty of NIST Measurement Results. *NIST technical Note 1297* (1994)
17. Joint Committee for Guides in Metrology, *Evaluation of measurement data Guide to the expression of uncertainty in measurement* (2008)

Chapter 5
Photonic, Plasmonic, Fluidic, and Luminescent Devices Based on New Polyfunctional Photo-Thermo-Refractive Glass

N. Nikonorov, V. Aseev, V. Dubrovin, A. Ignatiev, S. Ivanov, Y. Sgibnev and A. Sidorov

Abstract Fluoride Photo-Thermo-Refractive (PTR) glasses are very promising materials for recording Bragg gratings for different laser applications. Design and fabrication of novel chloride and bromide PTR glasses will be discussed. It was shown that various technologies as photo-thermo-induced crystallization, holograms recording, laser treatment, ion exchange, and chemical etching can be used for the cases of the fluoride, chloride and bromide PTR glasses, the so called polyfunctional. It is shown that polyfunctional PTR glasses can be used for the creation of novel optical elements and devices like holographic volume Bragg gratings, optical, luminescent and plasmonic waveguides, hollow structures, thermo-and biosensors, phosphors for LEDs, down-converters for solar cells have been designed and fabricated based on these new polyfunctional PTR glass.

5.1 Introduction

The current stage of development of optical, photonic and plasmonic devices calls for new and most likely miniature optical elements that cannot be fully implemented on the basis of traditional materials and technologies. Therefore, great attention is being paid worldwide to the development of novel optical materials.

Today, fluoride photo-thermo-refractive (PTR) glasses are among the most promising materials for several photonic applications. These glasses are widely used

N. Nikonorov (✉) · V. Aseev · V. Dubrovin · A. Ignatiev · S. Ivanov · Y. Sgibnev · A. Sidorov
Department of Optical Information Technologies and Materials, ITMO University, Saint Petersburg, Russia
e-mail: nikonorov@oi.ifmo.ru

© Springer Nature Switzerland AG 2018
P. A. Ribeiro and M. Raposo (eds.), *Optics, Photonics and Laser Technology*, Springer Series in Optical Sciences 218, https://doi.org/10.1007/978-3-319-98548-0_5

(a)   **(b)**

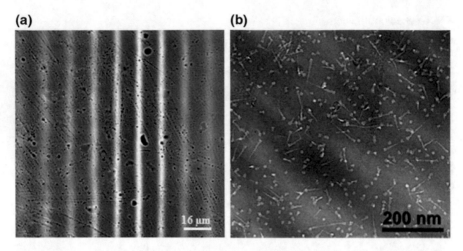

Fig. 5.1 Microscope image of the Bragg grating recorded in fluoride PTR glass (**a**) and TEM image of UV-exposed and heat-treated area of fluoride PTR glass, where NaF cone were grown (**b**)

for recording the volume Bragg gratings [1] and pictorial holograms [2], developing lasers and optical amplifiers [3, 4].

The standard PTR glass is a photosensitive multi-component sodium–zinc–aluminosilicate one containing fluorine (6 mol%) and bromine (0.5 mol%) and doped with small amounts of additives, as cerium, antimony and silver, that are responsible for the photo-thermo-induced precipitation of silver nanoparticles and sodium fluoride crystals [5–7]. The selective UV irradiation into the Ce^{3+} absorption band in the spectra of these glasses results in the formation of neutral silver molecular clusters that provide a broadband luminescence in the visible and NIR ranges [8, 9]. The subsequent heat treatment of UV-irradiated PTR glass near the glass transition temperature (Tg) induces the silver nanoparticle formation [7]. The thermal treatment of these glasses at temperatures above Tg leads to the growth of silver bromide shell on a silver nanoparticle [10] and then to the precipitation of a sodium fluoride cone on it [7, 11]. Image of a Bragg grating recorded in fluoride PTR glass and TEM image of NaF nanocrystals are shown in Fig. 5.1.

Bromine was shown to have a dramatic effect on the process of NaF crystal growth in fluoride PTR glasses [5]. The paper has demonstrated that the growth of sodium fluoride crystals is possible only in the presence of bromide additives in the PTR glass composition. A generalized NaF crystallization mechanism which consists of three stages has been proposed in [7, 12, 13].

At the first stage, the trivalent cerium ion donates an electron under the effect of the UV irradiation, thus increasing its own valency in accordance with the following reaction (5.1):

$$Ce^{3+} + h\nu \rightarrow e^- + \left[Ce^{3+}\right]^+ \tag{5.1}$$

Released photoelectrons can be trapped partially by silver ions (~20%) with subsequent neutral silver atoms and molecular clusters formation (Ag^0, Ag_2^0, Ag_2^+, Ag_3^{2+}) but most photoelectrons are trapped by antimony ions according to the following reaction (5.2):

$$e^- + Sb^{5+} \rightarrow \left[Sb^{5+}\right]^- \tag{5.2}$$

At the second stage, heat treatment at relatively low temperatures (300–450 °C) leads to the release of the trapped electrons from antimony (5.3) with further formation of silver molecular clusters and colloidal nanoparticles (5.4) according with:

$$\left[Sb^{5+}\right]^- \rightarrow Sb^{5+} + e^- \tag{5.3}$$

$$nAg^+ + ne^- \rightarrow nAg^0 \tag{5.4}$$

At the third stage, the heat treatment at temperatures above Tg results, first, in the growth of mixed silver bromide-sodium bromide shell on a silver nanoparticle and, further, in the coaxial growth of sodium fluoride crystalline phase on this shell.

It was also shown that UV irradiation and subsequent heat treatment of fluoride PTR glass induces the refractive index change only in the UV-irradiated area [10].

There is still some uncertainty in the origin of refractive index changes in PTR glasses and several presumable mechanisms have been proposed. Classically, this effect is assumed to be caused by differences in the refractive index between NaF crystalline phase (n ~ 1.33) that sediments in the UV-irradiated area and unexposed glass area (n ~ 1.49). Although a difference between the refractive indices of sodium fluoride and vitreous phases is rather high, the refractive index change in the UV-irradiated area does not exceed 1000 ppm [10, 14]. This can probably be due to the fact that, in addition to the NaF phase precipitation, there is the silver bromide shell with a high refractive index value (n ~ 2.3) on the silver nanoparticle. As shown in many sources (see, for example, [6, 15, 16]), the maximum surface plasmon resonance of silver nanoparticles in fluoride PTR glass shifts to greater wavelengths owing to the silver bromide shell growth.

Another possible mechanism of photo-thermo-induced refractive index change in PTR glasses has been proposed [17], which assumes that the transformation of Na and F distributed in PTR glass matrix into crystalline NaF (chemical changes) and structural relaxation process are not the main causes of photo-thermo-induced refractive index change but accounts it for high residual stresses around the NaF crystals. According to presented calculations, these stresses are the most important cause for photo-thermo-induced refractive index change in PTR glass.

By changing the type of halide (fluorine to bromine or chlorine) in the PTR glass composition it is possible to control a sign of the refractive index (RI) increment. In case of fluoride PTR glass, such treatment results in decrease of RI in UV irradiated area in comparison with unirradiated area. On the other hand, substitution of fluorine by chlorine and bromine in PTR glass, results in the precipitation of nano-crystalline phases of mixed silver and sodium chlorides, and silver bromide nanocrystals, respectively, in glass host and positive increment of RI (Δn up to 2200 ppm) [12, 18].

The fluoride, chloride and bromide PTR glasses developed in ITMO University (St. Petersburg, Russia) are very promising optical materials for optical, photonic and plasmonic applications [19]. The sizes of NaF, AgCl, and AgBr nanocrystals are relatively small (10–20 nm), that is why PTR glasses exhibit a rather low level of scattering. The PTR glasses can be successfully used for the fabrication of holographic optical elements (HOEs) that dramatically enhance properties of numerous laser systems and spectrometers. It presents high photosensitivity, high thermal stability of the recorded phase holograms, and high tolerance to optical and ionizing irradiation. Basic optical and spectral properties of PTR glasses are described in [6, 12, 18, 20]. The HOEs recorded in the PTR glass reveal high chemical stability, thermal, mechanical and optical strength, and from this point of view practically reveal no difference with the commercial optical glass BK7 (Schott). The optical and spectral parameters of the HOEs and GRIN-elements do not change after its multiple heating to the high temperature (500 °C). The most important advantages of having PTR glasses as optical medium are following: (i) high optical uniformity (the refraction index fluctuations across the glass volume have the scale of some 10^{-5}), (ii) reproducibility of its parameters during the starting glass synthesis and during the photo-thermo-induced crystallization, (iii) similarity to optical glass BK7, the PTR glass can be subjected to various methods of mechanical processing like grinding and polishing as well as various formation technologies like molding, aspheric surface production, and drawing fiber, (iv) one can fabricate the PTR glass both in laboratory (hundreds of grams) and in industrial (hundreds of kilograms) conditions with the use of simple and non-toxic technology. The chemical reagents, which are necessary for glass fabrication, are commercially available and not too expensive. One has also to note some features of PTR glasses, which are unusual for recording media. For example, PTR glasses can be subjected to ion-exchange technology, providing the possibility to fabricate the ion-exchanged optical or plasmonic waveguides and the surface strengthening to improve the mechanical strength, chemical stability, thermal and optical strength. PTR glass doped with rare earth ions reveals good laser characteristics [21, 22]. Recording the Bragg gratings in laser PTR glass opens up the chance to develop lasers with distributed feedback. The possibility to draw optical fibers from PTR glass and to record Bragg gratings in the fiber are shown [23].

Some characteristics of fluoride, chloride, and bromide PTR glasses and recorded volume Bragg gratings (VBGs) are listed in Table 5.1.

Table 5.1   Characteristics of fluoride, chloride, and bromide PTR glass

Parameter	Fluoride PTR glass	Chloride PTR glass	Bromide PTR glass
Transparency range, nm	350–3000	350–3000	350–3000
Photosensitivity spectral range, nm	280–350	280–350	280–350
Photosensitivity, mJ/cm^2	50	50	50
RI change, Δn	30×10^{-4}	22×10^{-4}	8×10^{-4}
RI modulation amplitude, δn	15×10^{-4}	11×10^{-4}	4×10^{-4}
Induced optical loss, cm^{-1}			
• visible range • near IR range	0.1 0.001	0.1 0.001	0.1 0.001
Space frequency, mm^{-1}	up to 5000	up to 5000	up to 5000
Diffraction efficiency, %	99	99	99
Hologram thickness, mm	0.1–10	0.1–10	0.1–10
Angular selectivity, ang. min	<0.005	<0.005	<0.005
Bandwidth FWHM, nm	<0.05	<0.05	<0.05
Size, mm	up to 25 × 25	up to 25 × 25	up to 25 × 25
VBGs are completely stable at temperature, °C	200	200	200

5.2   Holographic Optical Elements

5.2.1   *Volume Bragg Gratings for Laser Diodes*

The widespread use of semiconductor lasers is stimulated by a number of their advantages, such as high efficiency (75–80%) [24–26], small sizes, simplicity of operation, and relatively low cost. An important advantage of semiconductor lasers is the possibility of fabricating emitters operating at different wavelengths in the visible, near infrared, and mid-infrared spectral ranges. Apart from the above beneficial features, semiconductor lasers have certain drawbacks: their emission is quasi-monochromatic and spectrally unstable. This is caused by a number of factors as the broadening of the lasing spectrum, upon increase in the injection current stems from the fundamental aspects of charge-carrier transport and capture into the quantum-confined active

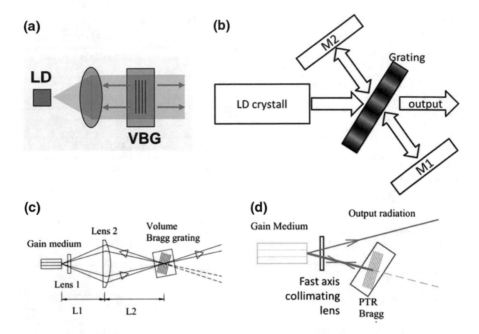

Fig. 5.2 Various designs of external cavity based on VBG. **a** simple case of reflecting VBG with collimation lens on the front facet of the LD crystal, **b** external cavity based on transmission VBG, **c** external cavity providing single more operation of the LD utilizing reflective VBG [26], **d** external cavity providing feedback for high order modes [26]

region and the multimode design of the laser cavity. A shift of the spectrum occurs as a result of heating the active region with increasing injection current, which causes a reduction in the band gap and, thus, the shift of the lasing spectrum to longer wavelengths. This problem can be solved by means of VBG recorded in the PTR glass. Due to the high spectral selectivity of recorded holograms, the implementation of such grating inside external laser diode cavity can significantly narrow the output spectra. This idea has been often used and proved its beneficial advantages. External cavity design based on the VBG can vary (see Fig. 5.2), as well as both reflecting or transmitting Bragg grating can be utilized [27].

Simplest implementation of VBG as an external cavity element is schematized in Fig. 5.2a, where radiation after collimating lens falls normally on the VBG element. Unfortunately, due to high divergence along the fast axis of the laser diode (LD) output radiation it is impossible to create reliable external cavity LD without collimation optics. Figure 5.2b shows typical design of external cavity using a transmission Bragg grating. Since the grating works back and forward and its diffraction efficiency has to be lower than 80% to efficiently couple output radiation there is need of an additional mirror in cavity setup to reduce power loss through non-diffracted radiation on backward cavity trip. Yet the position of mirrors can be changed, the number of output channels will stay the same. Figures 5.2c, d show cavities designs

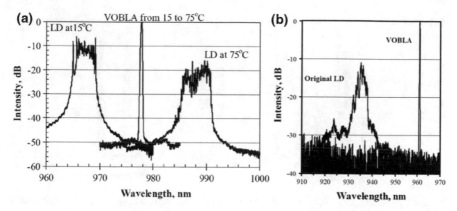

Fig. 5.3 Emission spectra of bare diode and volume Bragg laser (VOBLA) at different temperatures (**a**), emission spectra of single mode VOBLA (**b**) [27]

for coupling of the higher order modes of the LD. In this case, the reflecting VBG is placed in one arm of the mode and the other arm serves as output channel. Such a design requires high diffraction efficiency of the grating to provide maximum output performance unlike the other designs where diffraction efficiency has to be adjusted to maximize the laser performance. Cavity design based on higher order modes generation is suitable for wide stripes emitting diodes, since it was noticed that in this case they have greater gain.

Figure 5.3 represents temperature stability of the VBG stabilized laser diode in comparison with bare diode and spectral narrowing effect. Moreover, due to high transparency of HOE on PTR glass there are almost no losses in resulting output power.

Recent study of VBG based external cavity LD, shows that implementation of the grating inside such cavity significantly increases its selective properties. For instance, the grating analysed here [28] was recorded for the 1055 nm wavelength with Bragg angle of 30.7° and thickness of 1.3 mm. One can estimate its spectral selectivity as ~2 nm. The used cavity was the one depicted in Fig. 5.2a with transmitting VBG. Obtained emission spectra from such cavity design reveal laser output just on two longitudinal modes with separation of 100 pm and bandwidth of each one approximately 4–8 pm (Fig. 5.4).

Same as conventional ways of LD stabilization such as Littrow scheme and Litman-Metcalf configurations using standard diffraction gratings external cavities based on the VBG can provide tuning of the source output emission. By simple rotation of the grating, we can achieve tunability along all gain spectra of the semiconductor crystal, which can be really huge, up to 60 nm. An example of such tuning is shown on Fig. 5.5.

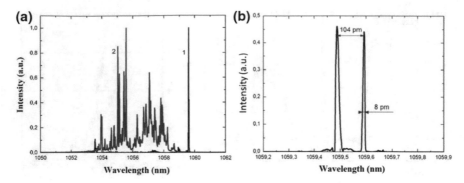

Fig. 5.4 Emission spectra of laser diode source with (1) and without (2) grating (**a**), detailed view
of the emission line (**b**) [28, 29]

Fig. 5.5 Emission spectra of
the external cavity laser
diode with different angles
of VBG

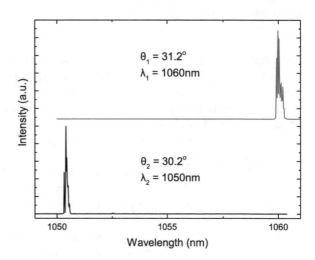

5.2.2 Imaging Holograms for Collimator Sight

Holographic collimator sights design follows from the development of the classical
collimator sights. Such a design provides greater transparency of the working aperture
in comparison with the classic collimator, in which aperture can be coloured due to
the use of anti-reflective (AR) coatings, as well as greater parallax suppression. As
these sights have open designs, both eyes can be aimed. A shooter can use peripheral
vision and engage more effectively. Also due to hologram properties, the scope will
be very resistant to various injuries and pollution. Since the hologram is recorded
over the entire aperture area, the sight remains in working condition with partial
pollution and damage. In addition, one of the best advantages to the conventional
scopes is the absence of the flare towards the target, which is crucial in combat.

Basic elements of a holographic sight are shown in Fig. 5.6. Operation principle of
holographic collimator sights can be briefly described as follows. Radiation from the

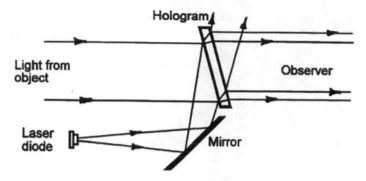

Fig. 5.6 Schematic image of first holographic collimator gun sight [30]

light source falls on the recorded hologram that creates an image of the recorded mark at the image plane. The hologram mimics the laser pointer effect with the difference that in this case there is no radiation towards a target. Obviously, the holographic element plays a key role, illuminated at a certain angle it reconstructs the reticle image at the image plane. Since the process of reconstruction of a hologram is very sensitive to the radiation wavelength and to the degree of collimation, additional optics of the formation of the illuminating beam is required.

Holographic elements used in existing sights are recorded on holographic plates Agfa 8E76 HD using a krypton laser (676 nm) [30]. However, dichromic gelatine has several disadvantages, such as high moisture sensitivity and the inability to get high hologram thickness [31]. The high value of refractive index modulation allows create a grating with diffraction efficiency greater than 90% at a thickness of several tens of microns. It should be noted that such high values of the refractive index (order of 10^{-2}) are not attainable in photorefractive crystals and glass. At the same time, VBGs cannot be recorded on the thicknesses of more than one hundred micrometres in this material, primarily due to the inhomogeneity of the thick layers of the medium.

On the other hand, high transparency of PTR glass in visible range (above 90% without AR coating) opens a new field of applications with strict requirements to transmission in observation channel, such as the collimator sight. For instance, typical transmission of a HOE with 90% diffraction efficiency recorded on a PTR glass is shown in Fig. 5.7.

Application of PTR glasses can solve problem of mark image stabilization, which is necessary due to the instability of laser diode source used in such scopes. This problem is to date solved by addition in the optical scheme of achromatizing diffraction elements such as additional thin gratings, complex two-cavity mirrors or compound objectives. Wavelength shifts, caused by changes in laser diode temperature, can be eliminated through spectral selectivity of a thick hologram recorded on a PTR glass. While the central wavelength of laser diode shifts, recorded hologram continues to reconstruct image of mark on proper angle—thus maintaining the position of mark in target plane.

Fig. 5.7 Transmission of the
HOE in the visible range

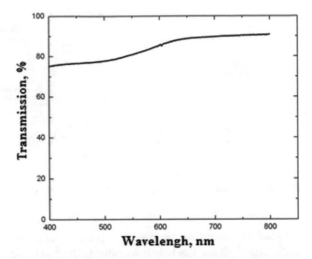

Fig. 5.8 Photo of a
reconstructed image of a
volume holographic mark
[29]

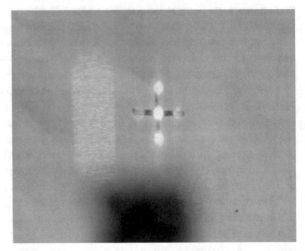

   Moreover, redistribution of energy in diode output spectrum leads to insignificant
lowering of intensity of the mark that can be easily levelled by diode power output
adjustment. Since the diffraction efficiency of PTR glass holograms can achieve
values of 99%, intensity required for mark observation is rather low. Worth to note
that up to date materials used for mark recording are vulnerable to external factors
such as moisture and mechanical damage, which leads to the need of additional
cover for holograms. With the use of PTR glasses, since they are highly resistant to
external factors, there is no need of additional protection of the observation channel.
The observable image of holographic mark recorded in a PTR glass is shown in
Fig. 5.8. This holographic element can be directly installed in the optical design of
the collimator sight shown at the Fig. 5.6.

5.3   Optical Amplifiers and Lasers Based on PTR Glass

5.3.1   *Optical Amplifiers*

Optical amplification has been demonstrated in a PTR glass doped with rare earth ions (Er^{3+} and Yb^{3+}) laser assembly. The concentration of ytterbium ions was 17.8×10^{20} cm^{-3} and the concentration of erbium ions was varied from 0 to 2.26×10^{20} cm^{-3}. Results showed that the introducing rare earth ions into the virgin PTR glass did not change its photosensitive properties. The experimental spectral dependences of the gain/loss coefficient for various pump power are shown in Fig. 5.9.

As the pumping power increases, the absorption spectrum transforms into a gain spectrum. The gain coefficient at wavelength of 1.55 μm was close to commercial Yb-Er silicate glass and achieved $g = 0.016$ cm^{-1}. The gain is obtained on the samples with a minimum erbium-ion concentration $N_{Er} = 0.26 \times 10^{20}$ cm^{-3} and $N_{Er} = 0.56 \times 10^{20}$ cm^{-3}. Increasing the erbium-ion concentration reduces the gain coefficient (Fig. 5.10).

Today the highly concentrated ytterbium-erbium glasses are used for fabrication of space-saving erbium doped waveguide amplifiers (EDWAs) working at 1.5 microns wavelength. The architecture of such space-saving EDWAs includes itself three key elements which are located on one substrate: (i) wavelength division multiplexing WDM element for combining pumping channel (0.98 μm) with signal one (1.53–1.56 μm) consist, (ii) waveguide amplifier, and (iii) Bragg grating for flattening the gain spectrum. Such packaging is named hybrid integration of different elements made from different materials. The use of different materials with different thermal expansion coefficients is a bottleneck in such approach. The combination of ion-exchangeable, laser and holographic properties into the laser polyfunctional PTR glass substrate allowed us to suggest a monolithic integration of WDM, amplifier

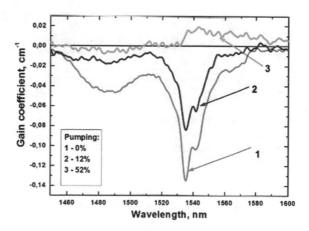

Fig. 5.9 Gain/loss spectrum of PTR glass ceramics with various pump power. $N_{Er} = 0.26 \times 10^{20}$ cm^{-3}, $N_{Yb} = 17.6 \times 10^{20}$ cm^{-3} [29]

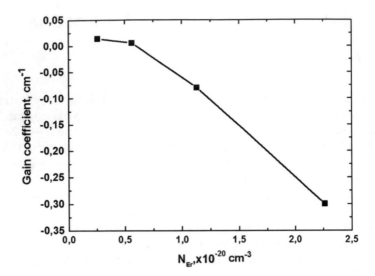

Fig. 5.10 The concentration dependence of the gain in the PTR glass ceramics. $N_{Yb} = 17.6 \times 10^{20}$ cm^{-3}. Pump power 590 mW

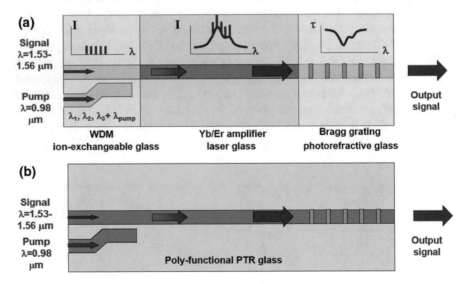

Fig. 5.11 Schematic description of a hybrid integration (**a**) of three elements: (i) WDM, (ii) amplifier channel and (iii) Bragg grating playing a role of spectral flattening filter that made with the use of three materials and monolithic integration (**b**) of these elements made with the use of polyfunctional PTR glass

channel and Bragg grating. Using the PTR material-substrate together in the mono-lithic approach is more attractive in comparison with hybrid integration (Fig. 5.11).

5.3.2 Distributed-Feedback/Distributed Bragg Reflector Lasers

The concept of distributed-feedback (DFB) lasers was originally demonstrated in 1971 [32], where the 1st laser output was achieved from a gelatin film on a glass substrate (Fig. 5.12a). The feedback of such a laser is provided by coherent Bragg scattering, from a periodic spatial variation of the gain medium refractive index or the gain itself. Two years later [33] generation on a similar structure in the GaAs (Fig. 5.12b) at liquid nitrogen temperatures was demonstrated. Benefits of such laser design are pretty obvious as Bragg grating acts as a selective mirror with very narrow reflection bandwidth and thus provide narrow spectra output emission. Since then DFB lasers had a great development but with no results concerning the creation of solid state DFB lasers.

Doping PTR glasses with rare earth elements provides access to the construction of DFB and distributed Bragg reflector (DBR) lasers since such medium possess both laser and holographic properties. Recently, the first results on laser action on PTR glasses have been reported by Sato et al. [34] and Ivanov et al. [35]. In the first work [34] a glass sample with concentration of Nd ions about $N = 1.38 \times 10^{20}$, which is similar to that used in commercially manufactured Nd:YAG crystals, and 1.93 mm thickness was used. Despite lack of AR coatings on the active element surfaces it showed good performance with overall slope efficiency of 24% (Fig. 5.13a) and 110mW output power. Yet the correct estimation of round trip losses (L) was not made in [34], which is general parameter of an active medium that affects overall laser performance. The measured value of L was rather large ~2.6% but authors connected it with the absence of AR coatings. Later on, generation on heavily Nd-doped PTR glass has been achieved [35], using a 2.5×10^{20} concentration of Nd^{3+} ions in prepared samples, reached value of that was twice from the one of previous study. Unlike in the previous work, authors used AR coatings and a 9 mm long sample to maximize the pump absorption. Our calculations show that PTR glass itself, due to outstanding homogeneity, possesses a relatively low round trip losses (0.26%), which is comparable with commercially fabricated Nd:YAG crystals. However, increase in Nd^{3+} concentration negatively affects lasing slope efficiency of PTR glasses, since with the output coupler of 1% only 16% slope efficiency has been achieved (Fig. 5.13b).

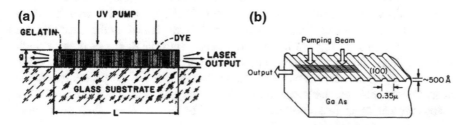

Fig. 5.12 Schematic image of DFB laser on polymer (**a**) [32] and DFB laser on semiconductor GaAs (**b**) [33]

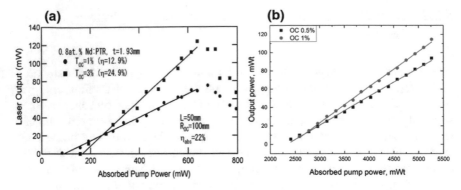

Fig. 5.13 Laser action of Nd-doped PTR glass with concentration of ~1.38 × 10²⁰ (**a**) [34] and heavily doped N = 2.5 × 10²⁰ (**b**) [35]

Yet, it is not clear why with definitely low round trip losses one could not reach 24% slope efficiency of the previous study.

Further investigations made towards DFB/DBR action on PTR glass showed that recording of the grating inside PTR glass does not affect the lasing properties. For instance, laser action in DFB/DBR configuration on Nd and Yb doped PTR glass was demonstrated [36]. In this experiments glass samples with 10–15 mm length with 2% of RE ions content were used. Both samples had VBG recorded inside the volume and basically DFB and DBR configurations differences as in the position of the dichroic mirror (Fig. 5.14), that is sort of inaccurate. Still, the authors achieved output radiation from both setups and on both types of PTR glasses (Nd and Yb ones). Nd laser was operating in pulsed regime DBR configuration and had a narrow emission line with slope efficiency of 15%, yet there was no information about pulse to pulse stability of output wavelength. The observed emission spectra shows narrow line with 30 pm bandwidth, but it is unclear if it is single mode operation or if there are several longitudinal modes since this observation was limited to spectral analyzers bandwidth. It was also noticed that with increase in pump power a red shift in emission wavelength occurs. This effect was found for both setups and overall value of the wavelength shift for Nd glass is estimated to be 350 pm.

Concerning to the Yb doped glass it is seen to perform in a CW regime using DFB configuration. A 10–13% slope efficiency was achieved, depending on the output direction. As well as in Nd laser it presents a red shift of operation wavelength of about 150 pm. The wavelength drift can be mainly explained by the glass thermal expansion which leads to a grating period change. Since PTR glasses present small refractive index changes with temperature, dn/dT < 1 ppm/K, and thermal expansion coefficient of 9.5 ppm/K, this problem can be circumvented by proper cooling of the glass. One can also see that slope the efficiency drops in described experiment, but it stays unclear which factor will be responsible for this behavior, either increase of RE concentration or grating recording itself.

Fig. 5.14 Experimental setups of DBR (**a**) and DFB (**b**) lasers 1—dichroic mirror with high transmission for pumping radiation and HR for signal at normal incidence (**a**) or at 45°, 2—lens, 3—Nd or Yb doped PTR glass slab with a 99% diffraction efficiency Bragg mirror

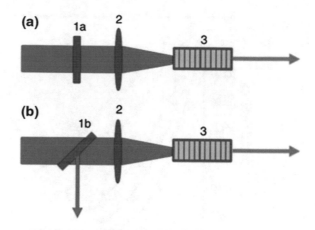

5.4 Optical Waveguides in PTR Glasses

The typical PTR glasses contains about 15 mol% of sodium oxide. This allows to consider the PTR glass as ion-exchangeable medium and use an ion-exchange (IE) technology for optical waveguides fabrication.

Glass IE technology was established in 70s and is nowadays straightforward towards the formation of optical waveguides and glassware mechanical strengthening [37–39]. Many advantages can be found in ion-exchanged waveguides on glass substrates, as for example quite low optical losses, excellent optical fibers compatibility, and both good mechanical and thermal stabilities. The IE waveguides process fabrication is simple and low-cost, as it does not require sophisticated instrumentation. Thus, simultaneously treatment of many samples can be easily carried out, enabling mass production and providing the required flexibility towards industrial processing. In addition, the refractive index profiles formation mechanisms in IE waveguide structures have been extensively investigated, in such a way that detailed crucial information can be easily found in literature [40–43].

For example, the low-temperature replacement of sodium ions in a glass by silver ions from a salt melt ($Na^+_{glass} \leftrightarrow Ag^+_{salt}$) is known to give rise to an increase in refractive index. This behavior is coming from the polarizabilities differences of the exchanging cations. For waveguides formed by the replacement of glass sodium ions by other alkali cations, as potassium, rubidium or cesium, the obtained refractive index profiles not only coming from differences in the polarizabilities of ions under exchange but also due to the photoelastic effect, emerging at the expenses of the compressive mechanical stresses in the waveguide plane [38]. The physico-chemical properties of a glass surface significantly change due to these stresses. Particularly, the exchange of alkali cations having different ionic radii such as $Na^+_{glass} \leftrightarrow K^+_{salt}$, causes increase in the microhardness, mechanical, thermal and optical durability of glass surface layers [44]. Under this compliance, regardless of the waveguide layer formation, the low-temperature IE can substantially enhance the glass surface

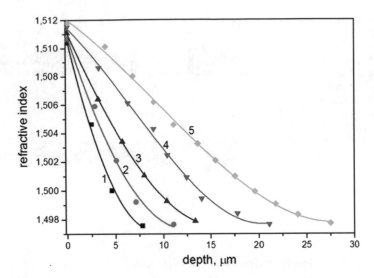

Fig. 5.15 Refractive index profiles of waveguides formed by the $Na^+ \leftrightarrow Ag^+$ IE. T = 350 °C and duration is (1) 15 min, (2) 30 min, (3) 1 h, (4) 2 h, and (5) 3 h

physico-chemical properties [45–47]. IE in PTR glasses was first carried out by Sgibnev et al. [48], which investigated the influence of PTR glass composition on the absorption spectra after the silver IE.

The optical waveguides in PTR glasses under analysis here have been prepared with the use of exchange of $Na^+_{glass} \leftrightarrow Ag^+, K^+, Rb^+, Cs^+_{salt}$ from melt of the corresponding nitrate ($0.1AgNO_3/99.9NaNO_3$ mol% nitrate mixture used for the silver IE). For $Na^+_{glass} \leftrightarrow Ag^+_{salt}$ optical waveguides no birefringence has been observed, observation that is consistent with the similarity of ionic radii of sodium (0.98 Å) and silver (1.15 Å). In graph of Fig. 5.15, where the waveguides refractive index profiles are plotted for different IE times, an increase from 15 min to 3 h leads to an increase in the waveguide depth up to 27 μm. The area under the curves is also seen to increase, indicating the increase in silver concentration in the surface layer. At the same time, the refractive index increment at the glass surface does not significantly change. This indicates a complete ion exchange taking place at the glass surface, where the silver-to-sodium ratio reaches the equilibrium magnitude. The maximum refractive index increment Δn for the Ag^+-waveguides in the experiments reached 0.0141 (Fig. 5.15).

The absorption UV edge presents a long-wavelength shift after the silver IE, in such a way this shift being known [49] it can be related with the absorption of diffusing silver ions which is characterized by an absorption band centered at 225 nm. There are no absorption bands in the visible or NIR range after the silver IE, allowing the use in optical waveguides formed in PTR glass in the wide optical range [29]. In case of $Na^+_{glass} \leftrightarrow Ag^+_{salt}$ IE the silver ions can be transformed into luminescent silver molecular clusters or plasmonic metallic nanoparticles by subsequent thermal treat-

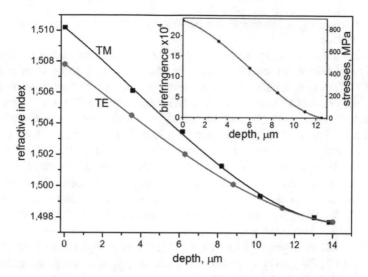

Fig. 5.16 Refractive index profiles of K$^+$-waveguides for the TE and TM polarizations. The IE temperature and duration were 350 °C and 6 h respectively. The inset shows the birefringence

ment at temperatures 400–500 °C. Such treatment allows to fabricate luminescent or plasmonic waveguides with a very strong surface plasmon resonance that can be used for chemical and biological sensors [50, 51].

Along the ion exchange $Na^+_{glass} \leftrightarrow (K^+, Rb^+,$ or $Cs^+)_{salt}$, both differences in the polarizabilities and ionic radii from K$^+$ (1.33 A°), Rb$^+$ (1.52 A°) to Cs$^+$ (1.67 A°) [52] are expected to influence the refractive index changes.

The refractive index profiles of K$^+$-waveguides obtained at IE temperature of 350 °C and duration of 6 h, are plotted in graph of Fig. 5.16 for both TE and TM polarizations. The birefringence magnitude at the glass surface, $\delta = n^{TM} - n^{TE}$, attains 0.0024 decreasing over the waveguide layer depth.

One can relate the magnitudes of compressive stresses σ and birefringence δ in a waveguide are interrelated by using the expression:

$$\sigma = \delta/B = (n^{TM} - n^{TE})/B, \qquad (5.5)$$

where n^{TM} and n^{TE} are the waveguide refractive indices respectively for the TM and TE polarizations, and B is the stress optical coefficient. The birefringence profile and corresponding stresses are displayed in Fig. 5.16 inset in terms of waveguide depth. Results show that high compressive stresses of 880 MPa can be attained at the PTR glass surface. Experimental data showed an increase in Vickers microhardness of PTR glass from 554 MPa before the potassium IE to 655 MPa after the IE carried out at 350 °C for 6 h. The maximum refractive index increment attainable for the K$^+$-waveguides is 0.0107. By increasing IE temperature from 350 to 420 °C a decrease in the refractive index increment can be observed, the behavior is associated with stress relaxation.

As rubidium ions are larger in size than the sodium ones they will present smaller mobility and diffusion coefficients in glass when compared to those of potassium ions. For this reason, longer periods were used to prepare the single and two-mode waveguides, 9 h and 18 h respectively, at the temperature of 350 °C. The maximum waveguide characteristics obtained for the $Na_{glass}^+ \leftrightarrow Rb_{salt}^+$ IE at 350 °C were refractive index increment of $\Delta n = 0.025$, for 18 h IE and birefringence of $\delta = 0.0035$, for 9 h IE. A waveguide depth of ~12 μm was attained rubidium IE at 420 °C for 18 h. The high birefringence value attained corresponds to stresses of about 1.2–1.3 GPa.

With respect to cesium ions which present the greatest ionic radius and, consequently, the lowest mobility in the glass, the respective Cs^+-waveguides prepared at $T = 420$ °C for 18 h, presented a refractive index increment of $\Delta n = 0.0512$, birefringence of $\delta = 0.002$, and a 3 μm depth.

With respect to absorption spectra, no significant changes have been observed when using K^+, Rb^+, and Cs^+ as diffusants in a wide spectral range from the UV to the IR in contrast to the Ag^+-waveguides.

The optical losses of the ion-exchanged waveguides do not exceed 0.5 dB/cm at 633 nm. Thus, the ion-exchangeable properties of the PTR glass are comparable with ones of commercial optical glass of BK7 and BGG31 used for optical waveguides fabrication [53].

5.5   Photoetchable PTR Glass for Microfluidic, MEMS and MOEMS Devices

The technology of chemical etching the fluoride PTR glasses has already been developed [54]. The rate of etching the crystalline phase is much higher than one for the glass host by factors of 6–10, depending on HF concentration in the aqueous solution [54]. The dependences of etched layer thickness for fluoride PTR glass and glass ceramics on duration of chemical etching in the 3 N HF solution are shown in Fig. 5.17. It is obvious that glass ceramics presents higher etching rate than the fluoride PTR glass. So, the fluoride PTR glass can be compared with well-known photoceramics as Foturan™ (Schott, Germany), Fotoform™ (Corning, USA) and PEG3™ (Hoya, Japan) with etching rates of 15–20 between the crystalline phase and the glass host, that have successfully been used for the fabrication of 3D hollow microstructures, microfluidic devices, micro-total analysis systems (μTAS), micro-electro-mechanical systems (MEMS) and micro-opto-electro-mechanical systems (MOEMS).

At the same time holographic properties of the Foturan™ and PEG3™ [55] greatly concede the PTR glass ones.

To demonstrate photostructurable properties of the PTR glass the Russian abbreviation of the ITMO University was chemically etched (Fig. 5.18).

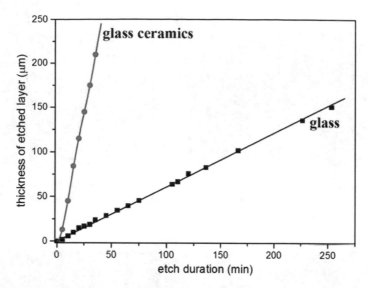

Fig. 5.17 Etched layer thickness versus the duration of chemical etching in the 3 N HF solution for PTR glass and PTR glass ceramics obtained by heat-treating the UV irradiated PTR glass at 500 °C for 10 h

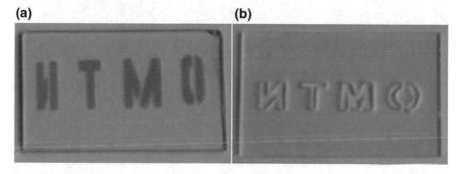

Fig. 5.18 UV irradiated through mask and heat-treated fluoride PTR glass (**a**). Subsequently chemically etched PTR glass (**b**) [29]

It should be noted that combination of photoetchable and ion exchangeable properties of the PTR glass could open prospects for developing new microfluidic and plasmonic devices. With respect to this it was shown that silver IE can impart the hydrophobic properties to glass surface [56]. For this reason, it is possible to improve the chemical durability, microhardness and hydrophilicity of microfluidic channels formed in the bulk of PTR glass substrate. As mentioned above, silver ions in the ion exchange layer can be transformed into either luminescent silver molecular clusters or into plasmonic silver nanoparticles, allowing to develop integrated microfluidic-plasmonic sensors.

(a)   (b)

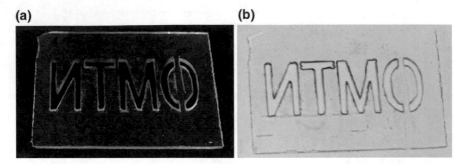

Fig. 5.19 Luminescent silver clusters (**a**) and plasmonic silver nanoparticles (**b**) formed by silver ion exchange and subsequent heat treatment inside of the hollow structures in volume of the fluoride PTR glass

(a)   (b)

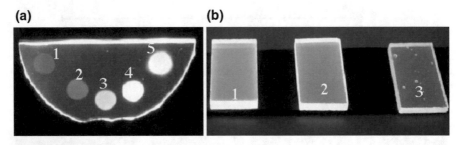

Fig. 5.20 Photos of chloride PTR glass luminesce. **a** Photo of PTR glass containing 1 mol% Cl after UV irradiation with various doses. The exposure duration (sec) that sets a dose is (1)—0.5 s, (2)—1 s, (3)—5 s, (4)—50 s, (2)—500 s. **b** Photo of UV irradiated and heat treated (1 h 400 °C) chloride PTR glass differing in chlorine concentration. The chlorine concentrations (mol%) being (1)—0, (2)—1, (3)—2

For example, we realized luminescent clusters and plasmonic nanoparticles inside of the hollow structures in volume of the fluoride PTR glass by ion exchange of Na^+_{glass} $\leftrightarrow Ag^+_{salt}$ and following thermal treatment (Fig. 5.19).

5.6 Phosphors Based on PTR Glass for LEDs and Down-Converters for Solar Cells

Luminescent silver molecular clusters (MCs) and complexes like "silver-halide" can be precipitated in the PTR glass by using ion-exchange technique or can be formed in the bulk of PTR glass host by UV radiation and subsequent heat treatment. These clusters and complexes have a broadband luminescence in the visible and NIR range under UV excitation [8, 57–60]. At the same time luminescent properties of the PTR glass have a strong dependence on glass composition and treatments (UV irradiation, heat treatment) parameters (Fig. 5.20).

An increase in luminescence intensity of the PTR glass after UV irradiation is caused by transform of silver ions Ag^+, and charged MCs Ag_n^+ ($n = 2$–4) to neutral state during UV irradiation and also by increasing neutral silver MCs concentration and size during subsequent heat treatment below the glass transition temperature. Luminescence intensity also depends on the concentration of halides in PTR glass [57]. Absolute quantum yield of silver clusters formed in PTR glass bulk reaches 50%. Formation of silver nanoparticles leads to decrease in luminescence intensity due to the decrease of MCs concentration and the appearance of plasmon resonance absorption band in the visible [57].

Thin layers with high concentration of luminescent silver molecular clusters can be formed by the silver ion exchange [8]. Spectra of the silver MCs formed in the bulk and surface layers of the PTR glass are quite similar (Fig. 5.21). The intensity and spectral region of MCs luminescence strongly depend on PTR glass composition as well as the temperature and duration of subsequent heat treatment. Luminescence efficiency grows with increasing of heat treatment temperature up to 450 °C, that can be associated with increasing concentration of silver molecular clusters. Heat treatment at temperature higher than the glass transition temperature (464 °C) results to formation of silver nanoparticles and luminescence quenching in the visible [8]. Concentration of reducing agents (Ce^{3+}, Sb^{3+}) existing in the PTR glass also significantly affects the luminescence efficiency. Increasing CeO_2 or Sb_2O_3 concentration in the PTR glass composition leads to luminescence quenching that can be caused by the formation of larger clusters that have weak luminescence in the visible and/or concentration quenching. Ion-exchanged PTR glasses with silver molecular clusters reveal white light emission and high absolute quantum yield (up to 63%) [61]. Figure 5.22 shows photo of silver MCs emission under UV excitation. Luminescent silver MCs formed in both planar waveguides and optical fibers can be used as concentrator for solar cells or temperature sensors.

As conclusion, the PTR glass can be successfully used as phosphors for white LEDs or down-converters for solar cells [62]. It should be pointed out, that these luminescent clusters and complexes could be formed in defined local spots of the PTR glass substrate by laser radiation. It allows creating complicated light architecture from luminescent centers.

5.7 PTR Glass for Optical Information Storage

The bright luminescence of neutral silver molecular clusters in PTR glass makes possible to record optical information in them by UV nanosecond laser irradiation. If the PTR glass was preliminary irradiated by the UV radiation into the absorption band of Ce^{3+} ions the UV nanosecond laser irradiation results in the silver clusters luminescence quenching (Fig. 5.23a). By not preliminary irradiate the PTR glass with the UV nanosecond laser, this results in neutral silver clusters luminescence appearance (Fig. 5.23b). The subsequent thermal treatment above the glass transition temperature results in the silver nanoparticles growth in the irradiated zones (Fig. 5.24b–d)

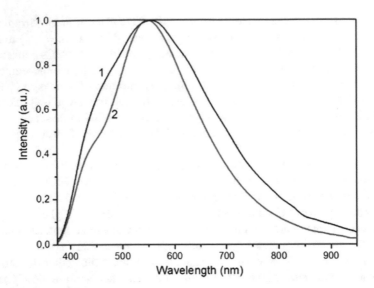

Fig. 5.21 Luminescence spectra of silver MCs formed in the bulk of preliminary UV irradiated (1) and in ion exchanged surface layer (2) of subsequently heat treated at 400 °C for 3 h PTR glass ($\lambda_{\text{ex}} = 360$ nm) [29]

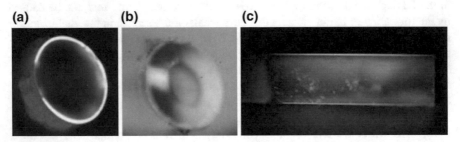

Fig. 5.22 Luminescence of silver MCs in fiber after 15 min (**a**) and 15 h (**b**) of the silver IE. Luminescence of silver MCs in the surface of the PTR glass after the silver IE (320 °C, 15 min) and subsequent heat treatment at 400 °C for 3 h (**c**). $\lambda_{\text{ex}} = 405$ nm

[57, 63–65]. By varying the glass composition the luminescent (Fig. 5.24b) or non-luminescent (Fig. 5.24c) silver nanoparticles can be grown in the irradiated zone. It can be seen from Fig. 5.24d that the formation of nanoparticles led to the local change of the glass color to yellow, reddish or brown. The described effects can be used for the optical information recording by the local change of glass luminescence or absorption.

Fig. 5.23 Negative and positive luminescent images in PTR glass plates irradiated through the mask by the UV nanosecond laser radiation with (**a**) and without (**b**) the preliminary irradiation by the UV mercury lamp [29]

5.8 Sensors Based on PTR Glasses

5.8.1 Luminescent Thermo-sensors

PTR glasses can be successfully used for the fabrication of luminescent thermo-sensors. The effect of temperature on the luminescence quenching of silver neutral MCs in PTR glass is illustrated in Fig. 5.25. As seen, the shape and location of the luminescence maximum on the wavelength scale remain intact under an increase in temperature (Fig. 5.25a), whereas the intensity of luminescence decreases [9]. In particular, an increase in temperature from −10 to +250 °C causes a decrease in the integrated intensity of luminescence by a factor of 25. Notably, the temperature dependence of the luminescence intensity shows no hysteresis and can be multiple times reproduced. Such properties of PTR glasses make them promising materials for the luminescent temperature sensors. The temperature dependence of luminescence intensity of PTR glass from −10 to +250 °C temperature range can be approximated quite satisfactorily (see Fig. 5.25b) by empirical function as follows:

$$I = 0.9(0.55 \exp(-T/25)) + 0.25 \exp(-T/150) \tag{5.6}$$

The former term is responsible for the low-temperature dependence while the second term is accounting for the high-temperature dependence. Complex nature of the function is associated with the presence of several types of silver molecular clusters in a glass.

(a)   (b)

(c)   (d)

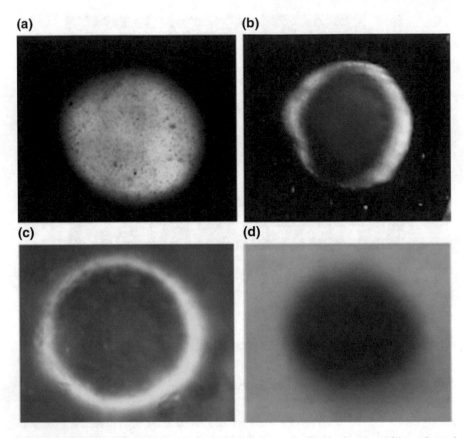

Fig. 5.24 Luminescence of irradiated zones after UV laser action (**a**, **b**, **c**). **a** without thermal treatment; **b**, **c** after thermal treatment above glass transition temperature for PTR glass without chlorine (**b**) and with chlorine (**c**). $\lambda_{ex} = 360$ nm. **d** image of irradiated zone after thermal treatment above the glass transition temperature

Also the fluoride PTR glasses doped with rare earth ions (for example, Er^{3+}) can be used for the fabrication of luminescent thermo-sensors operating in a wide temperature range (from -100 °C up to $+500$ °C). In this case, several effects can be used for measurement of temperature: (i) temperature deformation of profile of luminescence spectra, (ii) change of luminescence peaks of thermo-coupling levels, and (iii) temperature dependence of erbium luminescence life time [66]. Figure 5.26 shows emission spectra of Er-doped PTR glass at different temperatures.

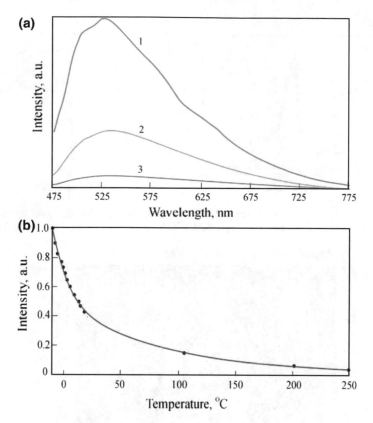

Fig. 5.25 The effect of temperature on the luminescence of PTR glass subjected to the UV irradiation and subsequent heat treatment. **a** The luminescence spectra recorded at temperatures of (1) −10 °C, (2) +25 °C and (3) +200 °C. **b** Normalized dependence of the integrated intensity of luminescence on temperature. $\lambda_{ex} = 360$ nm [39]

5.8.2 Plasmonic Sensors

Some technologies of control of concentration, size and shape of silver, metallic nanoparticles in bulk of the PTR glasses and on its surfaces (Fig. 5.27) have been developed for plasmonic sensors. The technologies allowed us to precipitate high concentration of silver nanoparticles with size of 10–100 nm in the PTR glass surface. The absorption coefficient of the plasmonic peak achieves more than 1000 cm^{-1}. These silver nanoparticles can have different shape: spherical, ellipsoidal and cubical. Some applications within this line have been demonstrated, as for example chemical and biological sensors based on localized surface plasmon resonance spectral position.

Silver nanoparticles can be formed on the surface of PTR glasses by the thermal treatment in reducing atmosphere or by the laser ablation or even evaporation of

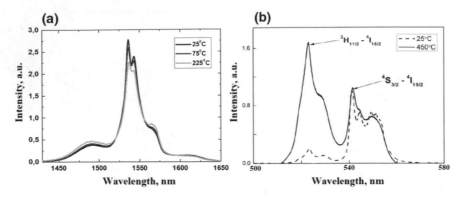

Fig. 5.26 Temperature deformation of profile of luminescence spectra (**a**), change of luminescence peaks of thermo-coupling levels (**b**). $N_{Er} = 2.26 \times 10^{20}$ cm^{-3}, $P_{pump} = 30$mW, $\lambda_{ex} = 975$ nm

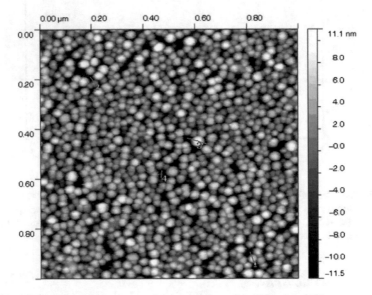

Fig. 5.27 AFM-image of silver nanoparticles on surface of the PTR glass [29]

the glass surface [63, 67–69]. For the last case, silver nanoparticles are covered by a 3–5 nm thick SiO$_2$ shell. Figure 5.28 shows the spectral position of the plasmon resonance peaks for the silver nanoparticles with (a) and without (b) dielectric shell on a PTR glass surface in air and in water. For the first case, the plasmon resonance spectral shift is 6 nm, while for the second case is 13 nm [70], which sufficient for the application in sensors elated with environment refraction index measurements.

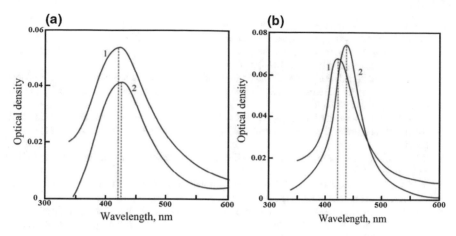

Fig. 5.28 Spectral position of the plasmon resonance peaks for the silver nanoparticles with (**a**) and without (**b**) dielectric shell on a PTR glass surface in air (1) and in water (2) [29]

5.9 Conclusions

Today one of the promising trends in photonics and plasmonics is miniaturization and integration optical elements on a single substrate. It becomes possible with using polyfunctional materials that allows applying various technologies and combine itself properties of several monofunctional materials. In this work new polyfunctional fluoride, chloride and bromide PTR glass were described. Photo-thermo-induced crystallization, ion exchange, and chemical etching technologies can be implemented in the new PTR glasses for various applications. Based on new fluoride, chloride and bromide PTR glass some examples of design and fabrication of novel photonic, plasmonic and microfluidic devices have been demonstrated as for example holographic volume Bragg gratings for diode lasers, optical amplifiers, lasers, optical and plasmonic waveguides, hollow structures, thermo- and biosensors, phosphors for LEDs and down-converters for solar cells.

Acknowledgements This work was supported by the Ministry of Education and Science of Russian Federation (Project 16.1651.2017/4.6).

References

1. A.L. Glebov, O. Mokhun, A. Rapaport, S. Vergnole, V. Smirnov, L.B. Glebov, Volume Bragg gratings as ultra-narrow and multiband optical filters, in *Proceedings of SPIE*, vol. 8428 (2012), p. 84280C–84280C–11
2. S.A. Ivanov, A.E. Angervaks, A.S. Shcheulin, Application of photo-thermo-refractive glass as a holographic medium for holographic collimator gun sights, in *Proceedings of SPIE*, vol. 9131 (2014), p. 91311B

3. L.B. Glebov, Photosensitive holographic glass—new approach to creation of high power lasers. Phys. Chem. Glasses: Eur. J. Glass Sci. Technol. Part B **48**(3), 123–128 (2007)

4. L.B. Glebov, V.I. Smirnov, C.M. Stickley, I.V. Ciapurin, New approach to robust optics for HEL systems. Proc. SPIE **4724**, 101–109 (2002)

5. N.V. Nikonorov, E.I. Panysheva, I.V. Tunimanova, A.V. Chukharev, Influence of glass composition on the refractive index change upon photothermoinduced crystallization. Glass Phys. Chem. **27**(3), 241–249 (2001)

6. L. Glebova, J. Lumeau, M. Klimov, E.D. Zanotto, L.B. Glebov, Role of bromine on the thermal and optical properties of photo-thermo-refractive glass. J. Non-Cryst. Solids **354**(2–9), 456–461 (2008)

7. S.D. Stookey, G.H. Beall, J.E. Pierson, Full-color photosensitive glass. J. Appl. Phys. **49**(10), 5114–5123 (1978)

8. Y.M. Sgibnev, N.V. Nikonorov, A.I. Ignatiev, Luminescence of silver clusters in ion-exchanged cerium-doped photo-thermo-refractive glasses. J. Lumin. **176**, 292–297 (2016)

9. V.D. Dubrovin, A.I. Ignatiev, N.V. Nikonorov, A.I. Sidorov, T.A. Shakhverdov, D.S. Agafonova, Luminescence of silver molecular clusters in photo-thermo-refractive glasses. Opt. Mater. **36**(4), 753–759 (2014)

10. L.B. Glebov, N.V. Nikonorov, E.I. Panysheva, G.T. Petrovskii, V.V. Savvin, I.V. Tunimanova, V.A. Tsekhomskii, New ways to use photosensitive glasses for recording volume phase holograms. Opt. Spectrosc. **73**(2), 237–241 (1992)

11. I. Dyamant, A.S. Abyzov, V.M. Fokin, E.D. Zanotto, J. Lumeau, L.N. Glebova, L.B. Glebov, Crystal nucleation and growth kinetics of NaF in photo-thermo-refractive glass. J. Non-Cryst. Solids **378**, 115–120 (2013)

12. N. Nikolay, I. Sergey, D. Victor, I. Alexander, New photo-thermo-refractive glasses for holographic optical elements: properties and applications, in *Holographic Materials and Optical Systems*, ed. by I. Naydenova (InTech, 2017)

13. I.M. Reviews, A review of the photo-thermal mechanism and crystallization of photo-thermo-refractive (PTR) glass, Dec 2016

14. T. Cardinal, O.M. Efimov, H.G. Francois-Saint-Cyr, L.B. Glebov, L.N. Glebova, V.I. Smirnov, Comparative study of photo-induced variations of X-ray diffraction and refractive index in photo-thermo-refractive glass. J. Non-Cryst. Solids **325**(1–3), 275–281 (2003)

15. J.J. Mock, D.R. Smith, S. Schultz, Local refractive index dependence of plasmon resonance spectra from individual nanoparticles. Nano Lett. 485–491 (2003)

16. N.V. Nikonorov, A.I. Sidorov, V.A. Tsekhomskiĭ, K.E. Lazareva, Effect of a dielectric shell of a silver nanoparticle on the spectral position of the plasmon resonance of the nanoparticle in photochromic glass. Opt. Spectrosc. **107**(5), 705–707 (2009)

17. J. Lumeau, L. Glebova, V. Golubkov, E.D. Zanotto, L.B. Glebov, Origin of crystallization-induced refractive index changes in photo-thermo-refractive glass. Opt. Mater. **32**(1), 139–146 (2009)

18. V.D. Dubrovin, A.I. Ignatiev, N.V. Nikonorov, Chloride photo-thermo-refractive glasses. Opt. Mater. Express **6**(5), 1701 (2016)

19. N. Nikonorov, V. Aseev, A. Ignatiev, A. Zlatov, New polyfunctional photo-thermo-refractive glasses for photonics applications, in *Technical Digest of 7th International Conference on Optics-photonics Design & Fabrication* (2010), pp. 209–210

20. A.M. Efimov, A.I. Ignatiev, N.V. Nikonorov, E.S. Postnikov, Quantitative UV-VIS spectroscopic studies of photo-thermo-refractive glasses. I. Intrinsic, bromine-related, and impurity-related UV absorption in photo-thermo-refractive glass matrices. J. Non-Cryst. Solids **357**(19–20), 3500–3512 (2011)

21. L. Glebova, J. Lumeau, L.B. Glebov, Photo-thermo-refractive glass co-doped with Nd3+ as a new laser medium. Opt. Mater. **33**(12), 1970–1974 (2011)

22. V.A. Aseev, N.V. Nikonorov, Spectroluminescence properties of photothermo-refractive nanoglass-ceramics doped with ytterbium and erbium ions. J. Opt. Technol. **75**(10), 676–681 (2008)

23. P. Hofmann, R. Amezcua-correa, E. Antonio-lopez, D. Ott, M. Segall, I. Divliansky, J. Lumeau, L. Glebova, L. Glebov, N. Peyghambarian, A. Schülzgen, Strong Bragg gratings in highly photosensitive photo-thermo-refractive-glass optical fiber. IEEE Photonics Technol. Lett. **25**(1), 25–28 (2013)
24. P. Crump, G. Erbert, H. Wenzel, C. Frevert, C.M. Schultz, K.-H. Hasler, R. Staske, B. Sumpf, A. Maaßdorf, F. Bugge, S. Knigge, G. Trankle, Efficient high-power laser diodes. IEEE J. Sel. Top. Quantum Electron. **19**(4) (2013)
25. I.S. Tarasov, High-power semiconductor separate-confinement double heterostructure lasers. Quantum Electron. **40**(8), 661–681 (2010)
26. N.A. Pikhtin, S.O. Slipchenko, Z.N. Sokolova, A.L. Stankevich, D.A. Vinokurov, I.S. Tarasov, Z.I. Alferov, 16 W continuous-wave output power from 100 μm-aperture laser with quantum well asymmetric heterostructure. Electron. Lett. **40**(22), 1413–1414 (2004)
27. G.B. Venus, A. Sevian, V.I. Smirnov, L.B. Glebov, High-brightness narrow-line laser diode source with volume Bragg-grating feedback. Proc. SPIE **5711**, 166–176 (2005)
28. S.A. Ivanov, N.V. Nikonorov, A.I. Ignat'ev, V.V. Zolotarev, Y.V. Lubyanskiy, N.A. Pikhtin, I.S. Tarasov, Narrowing of the emission spectra of high-power laser diodes with a volume Bragg grating recorded in photo-thermo-refractive glass. J. Semicond. **50**(6), 819–823 (2016)
29. Y.M. Sgibnev, N.V. Nikonorov, V.N. Vasilev, A.I. Ignatiev, Optical gradient waveguides in photo-thermo-refractive glass formed by ion exchange method. J. Lightwave Technol. **33**(17), 3730–3735 (2015)
30. J. Upatnieks, A.M. Tai, Development of the holographic sight, vol. 2968, pp. 272–281
31. D.J. DeBitetto, White-light viewing of surface holograms by simple dispersion compensation. Appl. Phys. Lett. **9**(12), 417–418 (1966)
32. H. Kogelnik, C.V. Shank, Stimulated emission in a periodic structure. Appl. Phys. Lett. **18**(4), 152–154 (1971)
33. M. Nakamura, H.W. Yen, A. Yariv, E. Garmire, S. Somekh, H.L. Garvin, Laser oscillation in epitaxial GaAs waveguides with corrugation feedback. Appl. Phys. Lett. **23**(5), 224–225 (1973)
34. Y. Sato, T. Taira, V. Smirnov, L. Glebova, L. Glebov, Continuous-wave diode-pumped laser action of Nd^{3+}-doped photo-thermo-refractive glass. Opt. Lett. **36**(12), 2257–2259 (2011)
35. S.A. Ivanov, V.F. Lebedev, A.I. Ignat'ev, N.V. Nikonorov, Laser action on neodymium heavily doped photo-thermo-refractive glass (2016), pp. 29–31
36. A. Ryasnyanskiy, N. Vorobiev, V. Smirnov, J. Lumeau, L. Glebova, O. Mokhun, C. Spiegelberg, M. Krainak, A. Glebov, L. Glebov, DBR and DFB lasers in neodymium- and ytterbium-doped photothermorefractive glasses. Opt. Lett. **39**(7), 2156–2159 (2014)
37. M.E. Nordberg, E.L. Mochel, H.M. Garfinkel, J.S. Olcott, Strengthening by ion exchange. J. Am. Ceram. Soc. **47**(5), 215–219 (1964)
38. S.S. Kistler, Stresses in glass produced by nonuniform exchange of monovalent ions. J. Am. Ceram. Soc. **45**(2), 59–68 (1962)
39. T. Izawa, H. Nakagome, Optical waveguide formed by electrically induced migration of ions in glass plates. Appl. Phys. Lett. **21**(12), 584–586 (1972)
40. A.N. Miliou, R. Srivastava, R.V. Ramaswamy, Modeling of the index change in K(+)-Na(+) ion-exchanged glass. Appl. Opt. **30**(6), 674–681 (1991)
41. R.V. Ramaswamy, R. Srivastava, Ion-exchanged glass waveguides: a review. J. Lightwave Technol. **6**(6), 984–1000 (1988)
42. J. Albert, G. Yip, Stress-induced index change for K+-Na+ ion exchange in glass. Electron. Lett. **23**(14), 737–738 (1987)
43. W.G. French, A.D. Pearson, Refractive index changes produced in glass by ion exchange. Am. Ceram. Soc. Bull. **49**(11) (1970)
44. N.V. Nikonorov, Influence of ion-exchange treatment on the physicochemical properties of glass and waveguide surfaces. Glass Phys. Chem. **25**(3), 207–232 (1999)
45. A.K. Varshneya, Chemical strengthening of glass: lessons learned and yet to be learned. Int. J. Appl. Glass Sci. **1**(2), 131–142 (2010)

46. A.K. Varshneya, The physics of chemical strengthening of glass: room for a new view. J. Non-Cryst. Solids **356**(44–49), 2289–2294 (2010)

47. S. Karlsson, B. Jonson, C. Stålhandske, The technology of chemical glass strengthening—a review. Glass Technol.: Eur. J. Glass Sci. Technol. Part A, **51**(2), 41–54, 2010

48. E.M. Sgibnev, A.I. Ignatiev, N.V. Nikonorov, A.M. Efimov, E.S. Postnikov, Effects of silver ion exchange and subsequent treatments on the UV-VIS spectra of silicate glasses. I. Undoped, CeO2-doped, and (CeO2+Sb2O3)-codoped photo-thermo-refractive matrix glasses. J. Non-Cryst. Solids **378**, 213–226 (2013)

49. A.M. Efimov, A.I. Ignatiev, N.V. Nikonorov, E.S. Postnikov, Photo-thermo-refractive glasses: effects of dopants on their ultraviolet absorption spectra. Int. J. Appl. Glass Sci. **6**(2), 109–127 (2015)

50. J. Homola, Present and future of surface plasmon resonance biosensors. Anal. Bioanal. Chem. **377**(3), 528–539 (2003)

51. K.A. Willets, R.P. Van Duyne, Localized surface plasmon resonance spectroscopy and sensing. Annu. Rev. Phys. Chem. **58**(1), 267–297 (2007)

52. T. Findakly, Glass waveguides by ion exchange: a review. Opt. Eng. **25**(2), 244–250 (1985)

53. N.V. Nikonorov, G.T. Petrovskii, Ion-exchanged glasses in integrated optics: the current state of research and prospects (a review). Glass Phys. Chem. **25**(1), 16–55 (1999)

54. Y. Sgibnev, N. Nikonorov, A. Ignatiev, V. Vasilyev, M. Sorokina, Photostructurable photo-thermo-refractive glass. Opt. Express **24**(5), 4563 (2016)

55. M. Kösters, H.-T. Hsieh, D. Psaltis, K. Buse, Holography in commercially available photoetch-able glasses. Appl. Opt. **44**(17), 3399–3402 (2005)

56. A. Razzaghi, M. Maleki, Y. Azizian-Kalandaragh, The influence of post-annealing treatment on the wettability of Ag+/Na+ ion-exchanged soda-lime glasses. Appl. Surf. Sci. **270**, 604–610 (2013)

57. V.D. Dubrovin, A.I. Ignat'ev, N.V. Nikonorov, A.I. Sidorov, Influence of halogenides on luminescence from silver molecular clusters in photothermorefractive glasses. Tech. Phys. **59**(5), 733–735 (2014)

58. K. Bourhis, A. Royon, G. Papon, M. Bellec, Y. Petit, L. Canioni, M. Dussauze, V. Rodriguez, L. Binet, D. Caurant, M. Treguer, J.J. Videau, T. Cardinal, Formation and thermo-assisted stabilization of luminescent silver clusters in photosensitive glasses. Mater. Res. Bull. **48**(4), 1637–1644 (2013)

59. A.S. Kuznetsov, V.K. Tikhomirov, V.V. Moshchalkov, UV-driven efficient white light genera-tion by Ag nanoclusters dispersed in glass host. Mater. Lett. **92**, 4–6 (2013)

60. A.I. Ignat'ev, N.V Nikonorov, A.I. Sidorov, T.A. Shakhverdov, Influence of UV irradiation and heat treatment on the luminescence of molecular silver clusters in photo-thermo-refractive glasses. Opt. Spectrosc. **114**(5), 769–774 (2013)

61. Y.M. Sgibnev, N.V. Nikonorov, A.I. Ignatiev, High efficient luminescence of silver clusters in ion-exchanged antimony-doped photo-thermo-refractive glasses: influence of antimony content and heat treatment parameters. J. Lumin. **188**, 172–179 (2017)

62. D.S. Agafonova, E.V. Kolobkova, I.A. Ignatiev, N.V. Nikonorov, T.A. Shakhverdov, P.S. Shir-shnev, A.I. Sidorov, V.N. Vasiliev, Luminescent glass fiber sensors for ultraviolet radiation detection by the spectral conversion. Opt. Eng. **54**(11), 117107 (2015)

63. D. Klyukin, V. Dubrovin, A. Pshenova, S. Putilin, T. Shakhverdov, A. Tsypkin, N. Nikonorov, A. Sidorov, Formation of luminescent and non-luminescent silver nanoparticles in silicate glasses by NIR femtosecond laser pulses and subsequent thermal treatment: the role of halogenides. Opt. Eng. **55**(6), in print (2016)

64. A.I. Ignatiev, D.A. Klyukin, V.S. Leontieva, N.V. Nikonorov, T.A. Shakhverdov, A.I. Sidorov, Formation of luminescent centers in photo-thermo-refractive silicate glasses under the action of UV laser nanosecond pulses. Opt. Mater. Express **5**(7), 1635 (2015)

65. D.A. Klyukin, A.I. Sidorov, A.I. Ignatiev, N.V. Nikonorov, Luminescence quenching and recov-ering in photo-thermo-refractive silver-ion doped glasses. Opt. Mater. **38**, 233–237 (2014)

66. V. Aseev, A. Abdrshin, E. Kolobkova, R. Nuryev, K. Moskaleva, N. Nikonorov, Thermal sensors based on ytterbium-erbium doped nano-glassceramics, in *Proceedings—10th International Conference on Laser and Fiber-Optical Networks Modeling, LFNM 2010* (2010), pp. 45–46

67. V.I. Egorov, A.I. Sidorov, A.V. Nashchekin, P.A. Obraztsov, P.N. Brunkov, Investigation of the morphological features of silver nanoparticles in the near-surface layers of glass when they are synthesized by heat treatment in water vapor. J. Opt. Technol. **80**(3), 174–178 (2013)
68. P.A. Obraztsov, A.V. Nashchekin, N.V. Nikonorov, A.I. Sidorov, A.V. Panfilova, P.N. Brunkov, Formation of silver nanoparticles on the silicate glass surface after ion exchange. Phys. Solid State **55**(6), 1272–1278 (2013)
69. V.I. Egorov, A.V. Nashchekin, A.I. Sidorov, Formation of an ensemble of silver nanoparticles in the process of surface evaporation of glass optical waveguides doped with silver ions by the radiation of a pulsed CO_2 laser. Quantum Electron. **45**(9), 858–862 (2015)
70. V.I. Egorov, A.I. Sidorov, Modelling of sensitivity of plasmon sensory elements based on silver nanoparticles obtained by laser evaporation and ablation. Opt. Spectrosc. **121**(1), 90–94 (2016)

Chapter 6
Mitigation of Speckle Noise in Optical Coherence Tomograms

Saba Adabi, Anne Clayton, Silvia Conforto, Ali Hojjat, Adrian G. Podoleanu and Mohammadreza Nasiriavanaki

Abstract Optical Coherence Tomography (OCT) is a promising high-resolution imaging technique that works based on low coherent interferometry. However, like other low coherent imaging modalities, OCT suffers from an artifact called, speckle. Speckle reduces the detectability of diagnostically relevant features in the tissue. Retinal optical coherence tomograms are of a great importance in detecting and diagnosing eye diseases. Different hardware or software based techniques are devised in literatures to mitigate speckle noise. The ultimate aim of any software-based despeckling technique is to suppress the noise part of speckle while preserves the information carrying portion of that. In this chapter, we reviewed the most prominent speckle reduction methods for OCT images to date and then present a novel and intelligent speckle reduction algorithm to reduce speckle in OCT images of retina, based on an ensemble framework of Multi-Layer Perceptron (MLP) neural networks.

6.1 Introduction

Optical imaging uses light to interrogate the morphological information of the compartments within a sample tissue. Optical imaging technologies can represent the internal structure of the sample across a range of spatial scales from micrometers to

S. Adabi · A. Clayton · M. Nasiriavanaki
Department of Biomedical Engineering, Wayne State University, Detroit, USA
e-mail: saba.adabi@uniroma3.edu

M. Nasiriavanaki
e-mail: mrn.avanaki@wayne.edu

S. Adabi · S. Conforto (✉)
Department of Applied Electronics, Roma Tre University, Rome, Italy
e-mail: silvia.conforto@uniroma3.edu

A. Hojjat
School of Physical Sciences, University of Kent, Canterbury, UK

A. G. Podoleanu
Applied Optics Group, University of Kent, Canterbury, UK

© Springer Nature Switzerland AG 2018
P. A. Ribeiro and M. Raposo (eds.), *Optics, Photonics and Laser Technology*, Springer
Series in Optical Sciences 218, https://doi.org/10.1007/978-3-319-98548-0_6

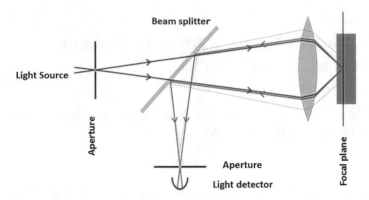

Fig. 6.1 Principle of a confocal microscopy

centimeters. Confocal microscopy (CM) and optical coherence tomography (OCT) are two modalities which work with the same principle. In CM [1], point illumination with a spatial pinhole is used in an optically conjugate plane in front of the detector to eliminate out-of-focus light (see Fig. 6.1).

In CM, the depth resolution (Δz) is inversely proportional to the square of the numerical aperture (*NA*) of the microscope objective lens (see (6.1)).

$$\Delta z = \frac{2n\lambda_0}{NA^2} \tag{6.1}$$

where λ_0 is the central wavelength of the source, n is the refractive index of the medium, d is the beam diameter at the objective lens, and f is the objective lens focal length [2]. NA can be approximated by:

$$NA \sim \frac{d}{2f} \tag{6.2}$$

Some parts of an OCT are borrowed from the confocal microscopy setup. The OCT principle of operation of is basically interferometry [3, 4]. Different interferometer configurations can be used to implement an OCT system [5]. Michelson interferometer is the most common one which is implemented using either a beam splitter or a directional coupler [5], together with a reference mirror, microscope objective lenses and a photodetector (Fig. 6.2). An OCT image is constructed based on the principle of time of flight [6]. The interferometry is used to magnify the very small time delay between the backscattered light returned from the sample and the reflected light from a reference mirror (Fig. 6.2). The basic components of an OCT system are a low coherent light source, to be able to have depth sectioning capability; a beam splitter to split light between two arms; a reference mirror, and some opto-electronic components such as objective lens and XY galvo-scanner [7]. The schematic of an OCT system is shown in Fig. 6.2.

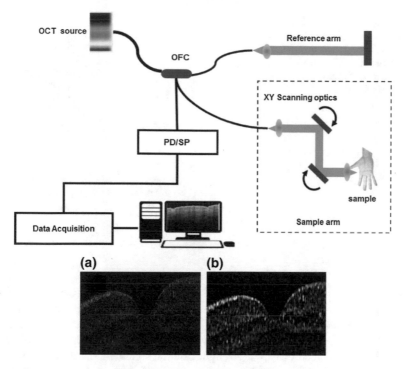

Fig. 6.2 Left: Simplified configuration of OCT systems, OCT source (Swept laser source or a broadband source), PD: Photodetector/SP: spectrometer, OFC: optical fiber coupler, L: Lenses, Right: **a** an original OCT image, **b** a despeckled OCT image

In this configuration, the optical source couples light into an optical fiber. The light is then split into two paths through the coupler/beam splitter, depending on the splitter ratio, a portion of light goes towards a sample arm and the rest towards a reference arm. The beams reflected from the reference and sample arms are recombined after passing through the coupler again. The recombined signal is collected by a photodetector. From interference theory, the photodetected signal is:

$$i(x) = k\left[R_s + R_r + 2\sqrt{(R_s + R_r)}\prod \cos\left(\frac{2\pi}{\lambda_0}\right)\right]\frac{P_0}{2} \qquad (6.3)$$

where k is the photodetector responsivity, R_s is the sample reflectivity, R_r is the reference mirror reflectivity, $\prod$ is the degree of polarization i.e. the degree of similarity of the orientation of the electric fields of the two optical beams, and P_0 the optical power at the object. Given that the optical path length for the sample and reference arms are I_s and I_r, respectively, the optical path difference (OPD) between the two optical paths in the two arms is $x = |I_s - I_r|$. In (6.3), the third term represents the interference signal, that is periodic and dependent on x and λ_0. Let us say that the optical source is broadband, such as a superluminescent diode (SLD). Each time the

Fig. 6.3 Relative orientation
of the axial scan (A-line), en
face scan (T-line),
longitudinal slice (B-scan)
and en face or transverse
slice (C-scan) [5]

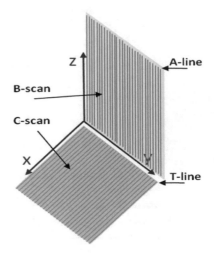

reference mirror is moved by an extra $\frac{\lambda_0}{2}$ (which determines a round trip OPD of λ_0), the photodetected signal exhibits a peak. A high pass filter is normally utilized to filter out the d.c. components of the signal and separate the a.c. component. The envelope of the OCT signal is retained by rectifying and filtering the signal amplified by the photodetector [8]. The strength of such signal versus the position of the reference mirror, gives a reflectivity profile, designated as axial scan (A-scan or A-line). A set of A-lines taken for different transversal positions of the incident beam on the sample, will form a cross section image a so called B-scan. If the collection of data is first performed transversally and then in depth, en face scan (T-line) and transverse slice (C-scans) are generated, respectively. This is the principle of a time domain (TD) OCT. The definition of A-line, B-scan and T-line, C-scan are illustrated in Fig. 6.3.

The interference between the wavefronts, coming back from the reference and sample arms, takes place only when the optical path difference is within the coherence length of the source. The coherence length, l_c, is defined as the length over which a wave maintains phase relations (6.4). The axial resolution, Δz, of an OCT system is determined as half of the coherence length of the source (6.5). The lateral resolution in the focal plane is defined by (6.6) [6, 5]. In the (6.4) to (6.6), λ_0 is the central wavelength of the optical source and $\Delta\lambda$ is the full width half maximum (FWHM) of the optical source spectrum. By blocking the reference arm and measuring the optical power in the photodetector versus imaging depth, a profile is obtained of which the FWHM is measured which is known as the confocal gate. There is also the coherence gate which is directly proportional to the coherence length of the optical source.

$$l_c = \frac{4 \ln 2 \lambda_0^2}{\pi \Delta\lambda} = 0.88 \frac{\lambda_0^2}{\Delta\lambda} \qquad (6.4)$$

$$\Delta z = \frac{l_c}{2} \tag{6.5}$$

$$\Delta x = \frac{0.61\lambda_0}{NA} \tag{6.6}$$

6.1.1 Optical Coherence Tomography

Optical coherence tomography (OCT) is a non-invasive, non-ionizing optical imaging technique which works based on low coherence interferometry [4]. The OCT image, is made of the transversal measurement over a biological sample of the magnitude and time delay of backscattered infrared returned light [9, 5]. It is able to provide high resolution images and moderate penetration depth, e.g., one to three millimeters. OCT is currently utilized in several medical and biomedical applications including dermatology [10–12], dentistry [13], oncology [14], and cardiology [15] in addition to its initial successes in ophthalmology [16]. Recently, OCT has been used as an optical biopsy method for differentiation among different tissues, e.g., healthy versus tumorous [17, 18]. Quantitative analysis of OCT images through extraction of optical properties has made OCT an even more powerful modality [19–21]. An OCT system is characterized by several parameters such as imaging speed, lateral and axial resolutions and penetration depth [22].

Optical coherence tomography is an advanced high resolution, non-invasive imaging modality suitable to obtain three-dimensional (3D) images of microstructures within tissues. Several modifications in both OCT hardware and software have already been undertaken in improving these capabilities. Yet, OCT images are still not free of artifacts [22], which are essentially arising from speckle noise, intensity decay and blurring [23, 24]. Speckle degrades the OCT images quality masking diagnostically relevant features, drawback which is common to other low coherent imaging modalities, [25, 26, 13, 4]. With respect to blurring it mainly causes depreciation of OCT images lateral resolution. Addressing these issues, OCT images need to be enhanced in order to deliver microscopic features of biological samples more effectively.

6.1.2 Speckle Reduction

In optical imaging systems, grainy texture on images is always imposed by speckle reducing both signal-to-noise ratio (SNR) and their contrast-to-noise ratio (CNR). Therefore, speckles impairs procedures towards image segmentation and pattern recognition that are used to extract, analyze, and recognize image features. In OCT imaging, if the central wavelength of the light source is equal to or larger than the compartments within the sample under investigation, the interference of the reflected light with different amplitudes and phases generates a grainy texture in the image called speckle. Speckle degrades the quality of OCT images, particularly the borders of cellular layers [25]. The probability density function (PDF) of the speckle has been approximated by Rayleigh distribution, or Rician distribution [27]. Speckle pattern is highly dependent to the microstructural content (size and density) of the sample being imaged. Due to this correlation, speckle is also known to carry some morphological information, thus is not appropriate to consider it as an image noise. This issue has made finding a suitable solution to reduce the speckle quite challenging [28]. The speckle reduction methods are categorized into two main categories: software based and hardware methods [29, 19, 25, 30–36].

6.1.2.1 Hardware-Based Despeckling Methods

The most common hardware-based speckle reduction method is compounding. In compounding techniques [37], partially de-correlated images acquired from stationary samples are averaged. The quantities to be averaged specify the compounding procedure. Some of the quantities used in compounding methods are backscattering angles, central wavelengths, polarizations, and displacements. These results in techniques referred to as angular compounding, frequency compounding, polarization compounding, and spatial compounding, respectively [38, 31, 39, 4]. For instance, in the spatial compounding method, the averaging quantity is the tissue or the imaging probe motion, which comes from the inherent imperfection of the scanners used in the configuration of the imaging system [40]. In a study, five different algorithms were used including averaging, random weighted averaging, random pixel selection and random pixel selection plus median filtering to average their partially correlated images obtained from the spatial compounding method (Fig. 6.4) [40]. Findings demonstrated that the random pixel selection plus median filtering method is an efficient, simple, and edge-preserving despeckling method compared to the common averaging method.

6.1.2.2 Software-Based Speckle Reduction Methods

Software based speckle reduction methods rely on a mathematical model of the speckle, and they can be classified into adaptive and non-adaptive filters. The former are implemented based upon the local first order statistics, such as mean and variance, while the latter are implemented based on the overall statistics in the image. Wiener filter is one of the most popular adaptive methods [41, 42]. Some of the non-adaptive algorithms are Kuwahara filter, Hybrid Median filter, Enhanced LEE filter (ELEE), Symmetric Nearest Neighborhood (SNN), thresholding with fuzzy logic [43, 44]. Wavelet based despeckling has been a successful non-adaptive despeckling method in which the image is decomposed into its wavelet bases, allowing to differentiate noise components through some signal processing [29, 8, 45, 46, 22]. Considering the importance of wavelet mother function in this method, Haar mother function has proven a fast and efficient solution, enabling speckle noise reduction without

Fig. 6.4 Results of spatial compounding algorithms performed on OCT images of larynx: **a** original image, enhanced image via **b** averaging, **c** median filtering, **d** random weighted averaging, **e** random pixel selection, **f** random pixel selection plus median filtering

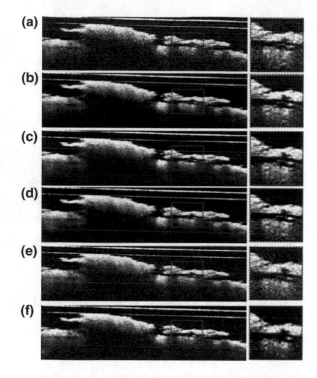

substantially diminishing contrast or spatial resolution in the image [47]. An artificial neural network based (ANN) methodology towards speckle reduction has been previously introduced [48–52]. In this methodology, speckle was modelled through a single noise parameter Rayleigh distribution for the entire image. The algorithm was successfully applied to Drosophila larvae OCT images, and it is believed to be worth of further improvements towards better efficiency. ANN offers an intelligent solution which reduces speckle while preserving the morphological information in the image. In this method, the speckle is first modelled. The utilized model follows Rayleigh distribution and is given by (6.7) [53, 54].

$$f(x_{i,j}) = \frac{x_{i,j} e^{\frac{-x_{i,j}^2}{2\sigma^2}}}{\sigma^2} \tag{6.7}$$

where $x_{i,j}$ is the image pixel and σ is the image noise variance (the so called noise parameter). A cascade forward back propagation ANN is then used to estimate an image noise parameter, after which a numerical solution to the inverse Rayleigh distribution function is carried out [49]. The OCT images of Drosophila heart before and after applying ANN based despeckling method are shown in Fig. 6.5.

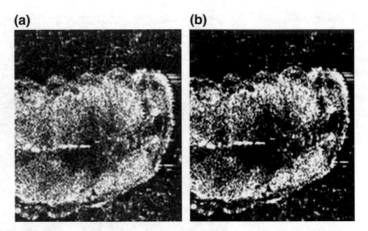

Fig. 6.5 C-scan images acquired from a Drosophila heart at depth of 300 μm. **a** Original image, **b** de-speckled image

6.2 Intelligent Speckle Reduction Method

6.2.1 Artificial Neural Network

An Artificial Neural Network (ANN) consists of an information processing paradigm based on the manner human biological nervous systems, i.e., brain, use to analyze information. Here the paradigm main component is the structure through which the processing of information is performed. This structure consists of a large number of neurons, i.e., highly interconnected processing components, that work constructively and coherently to solve specific problems. Similar to human, ANNs learn by examples to perform in a specific application such as pattern recognition or data classification. Mathematically, learning means adjustments to the synaptic connections between the neurons. ANNs have capability to analyze complicated data; they can be used to derive patterns and detect trends that are too complex to be analyzed by simple computer techniques. A trained ANN can be thought of as an "expert", in the given problem. This expert will then be able to predict a solution/answer to new situations. There are more advantages worth to be mentioned namely: (i) adaptive learning which is the ability to learn how to do tasks based on the data given for training or initial experience; (ii) self-organization consisting of creation of own organization or representation of the information received during learning; (iii) real time operation which means parallel computations capabilities; (iv) fault tolerance via redundant information coding, which is the partial destruction of a network leading to the corresponding performance degradation, although some network capabilities may be retained even with major network damage. A neuron can be thought in a simple way as a device having several inputs and a single output (Fig. 6.6). The neuron usually has two operational modes: training and test. In the training mode, the neuron is

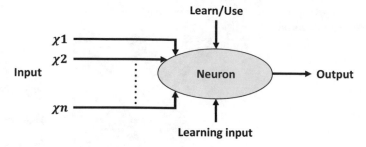

Fig. 6.6 A simple neuron representation

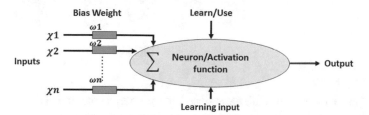

Fig. 6.7 A MCP neuron representation

trained to work/fire or not, for particular input patterns. In the using mode, the output is generated according to the input/output list taught. However, if the input is not found in the input/output list taught, the firing rule is used to determine whether to fire or not.

With a more complex neuron, more complicated tasks can be performed that cannot be already done in computers (Fig. 6.7). McCulloch and Pitts model (MCP) is the complex neuron. In this model, the inputs are 'weighted', and each input has a decision-making power depending on the weight of the input. The weight is a constant value that is multiplied by the input. The weighted inputs are then added together and compared to a pre-set threshold value. If the value is larger, the neuron fires, otherwise not.

Having neurons with weighted inputs and a threshold makes them a small sophisticated processing unit. In fact, by altering the weights and/or threshold of an MCP neuron, the neuron has ability to adapt to any problem. Several algorithms exist that reason the neuron to 'adapt', e.g., feedback networks (Fig. 6.8) that are frequently used in a pattern recognition problem [55, 56].

6.2.2   Speckle Reduction Method

In the followings, a new scheme is presented, in which the image is segmented into sections. Thereafter the MLP neural networks for different segments are used to esti-

Fig. 6.8 An example of a
simple feed forward network

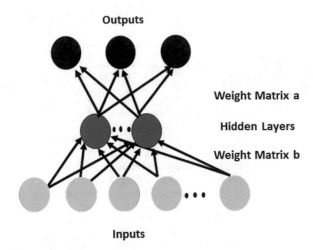

mate the noise parameter. With these steps and a numerical method, the segments and image were de-noised. Further processing was performed to eliminate the blocking artifact. The algorithm is carried out in two phases a follows. The training phase is the first one where a Rayleigh noisy image generator is used. For this purpose 10 × 10 pixels images were generated in MATLAB with sigma values (the single noise parameter used in Rayleigh function) ranging from 0 to 255 in steps of 0.05. This procedure is repeated 100 times for each sigma value, to generate numerous training noisy images.

The average, standard deviation, and median were calculated from each segment and its wavelet sub-bands for training. The frequency domain statistical knowledge of the image was calculated from the wavelet sub-band images. A combination of several MLP neural networks was used as neural network. The used algorithm flow chart is displayed in Fig. 6.9. The main components of this framework are three MLP networks and a combiner which is responsible for the averaging process. Each of the MLP networks is composed of 15 neurons in the input layer, 10 neurons in the hidden layer and one output neuron for the estimation of sigma parameter. The combiner accounts for averaging with L neurons in input layer, L neurons in hidden layer and one output neuron which can estimate the sigma parameter in an ensemble. In the case considered here L = 3. In order to demonstrate the advantages of the ensemble method over individual neural networks, let's consider a number L of trained MLP neural networks with outputs $y_i(\underline{x})$ for an input vector x. These estimate the sigma values using the ith MLP neural network with an error e_i with respect to the desired sigma parameter value, $h(\underline{x})$. Under these conditions, $y_i(\underline{x})$ can be written as:

$$y_i(\underline{x}) = h(\underline{x}) + e_i \tag{6.8}$$

In such a way that the sum of squared error for the network y_i can be calculated as follows:

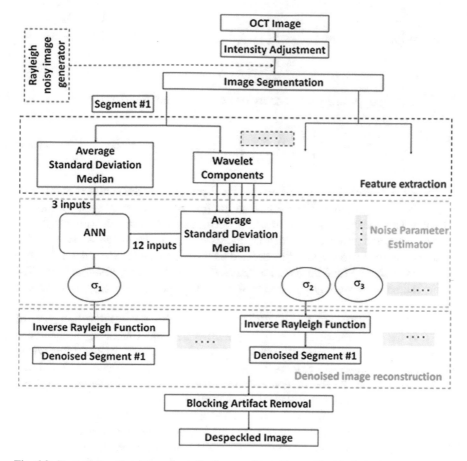

Fig. 6.9 Despeckling algorithm schematic diagram. Dotted boxes The blocks in the dotted box are used in the training phase

$$E_i = \xi\left(y_i(\underline{x}) - h(\underline{x})\right)^2 = \xi\left[e_i^2\right] \qquad (6.9)$$

where $\xi[.]$ stands for average or mean value expectation. So, the average error for the MLP networks acting individually can be expressed as:

$$E_{AV} = \frac{1}{L}\sum_{i=1}^{L} E_i = \frac{1}{L}\sum_{i=1}^{L} \xi\left[e_i^2\right] \qquad (6.10)$$

Committee prediction can then be obtained by averaging the y_i outputs. The error associated with this estimate will be:

$$E_{COM} = \left(y_{COM}(\underline{x}) - h(\underline{x})\right)^2 = \left[\left(\frac{1}{L}\sum_{i=1}^{L} y_i(\underline{x}) - h(\underline{x})\right)^2\right] = \xi\left[\left(\frac{1}{L}\sum_{i=1}^{L}[e_i]\right)^2\right]$$

(6.11)

Thus, using the Cauchy's inequality one can show that $E_{COM} \leq E_{AV}$.

$$E_{COM} = \xi\left[\left(\frac{1}{L}\sum_{i=1}^{L}[e_i]\right)^2\right] \leq \frac{1}{L}\sum_{i=1}^{L}\xi[e_i^2] = E_{AV}$$

(6.12)

The largest reliability in sigma estimation dispatched by the neural network took place when using a Daubechies 4 (db4) mother function. The neural network gathers, a total of 12 inputs: The transfer function, performance function, learning function, network size, and number of hidden layers, were experimentally chosen in order to attain optimum network reliability, defined as the percentage ratio of the difference between the expected and estimated sigma values over the expected sigma value. The sigma estimator network largest averaged reliability, measured over 20 runs, was of 99.3%.

The second phase will be testing. As shown in the despeckling flowchart of Fig. 6.9, the OCT image is initially segmented in terms of homogeneity. As in the training stage, the same pre-processing is applied to each image, after which the statistical features are extracted from each image segment and used as input of the neural network. The trained network is then used to estimate each segment Rayleigh noise parameter [10]. Thereafter the estimated sigma is used to numerically solve the inverse Rayleigh function for each segment. By placing together the noise model segments, a noise model image can be generated. The noise model image was deducted from the original image through a experimentally obtained scale factor. For the blocking artifact removal, statistical features were extracted from the original image, and the despeckled segments joined together, as follows from the method outlined in [57].

6.2.3   Experimental Results and Discussion

The retinal images were generated from a spectrometer-based Fourier-domain OCT system. The instrument layout is schematically shown in Fig. 6.10. The light from a two spectrally-shifted super-luminescent diodes (SLDs) of central wavelengths $\lambda_0 = 890$ nm and $\Delta\lambda = 150$ nm linewidths from Superlum Broadlighter D890), is coupled to the interferometer's arms through a fiber-based directional coupler (FDC). The sample arm comprises a scanning galvanic mirror (SX) together with a f-2f-f lens assembly specifically designed for retinal imaging.

Fig. 6.10 Spectrometer-based Fourier-domain optical coherence tomography system. SLD: super-luminescent diode with a central wavelength $\lambda 0 = 890$ nm and linewidth $\Delta\lambda = 150$ nm, L1–L3: Achromatic Lenses, CMOS: Linear pixel array (line camera), SX: galvo-scanning mirror, TG: diffraction grating, M1: flat mirror, PC: polarization controller, FDC: 80/20 fused fiber directional coupler, DC: dispersion compensating element

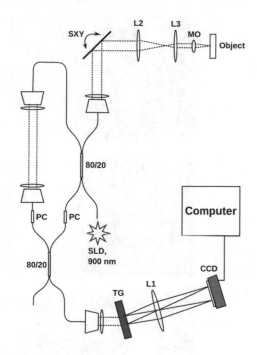

Table 6.1 Estimated sigma values corresponding to 8 segments (4 segments in each row) [53]

(a)		(b)		(c)	
Segment #	Sigma	Segment #	Sigma	Segment #	Sigma
1	112	1	98	1	75
2	56	2	64	2	64
3	21	3	51	3	56
4	41	4	45	4	68
5	131	5	68	5	28
6	145	6	20	6	30
7	162	7	39	7	27
8	143	8	81	8	19

The algorithm was tested with In vivo obtained eye B-scan OCT images. The OCT images obtained before and after the application of the de-speckling algorithm are displayed in Fig. 6.11.

As the number of segments in each image affects the despeckling efficiency, for edge sharpness improvement, towards more effective blocking artifact removal, the images were segmented into eight sub-images. The estimated sigma values for the segments within the images are displayed in Table 6.1a–c, each corresponding to three images I1, I2 and I3 of Fig. 6.11 respectively.

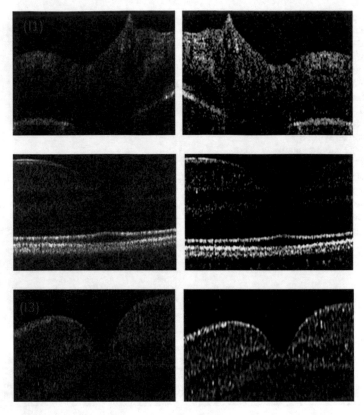

Fig. 6.11 Comparative presentation of (on the right) three original OCT test images (I1), (I2), (I3) and their respective denoised images (on the left)

I order to evaluate the images enhancement as a result of despeckling, the signal-to-noise ratio and the contrast-to-noise ratio (CNR) were evaluated as according with (6.8) and (6.9) respectively [58, 37].

$$SNR = 10 \log_{10}(\frac{(\max I)^2}{\sigma_b^2}) \qquad (6.8)$$

$$CNR = \frac{1}{R}\left(\sum_{r=1}^{R} \frac{(\mu_r - \mu_b)}{\sqrt{\sigma_r^2 + \sigma_b^2}}\right) \qquad (6.9)$$

where, max(I2) represents the maximum of squared intensity pixel values in a homogeneous region of interest within the linear magnitude image; μ_b and σ_b^2 and CNR is the mean variance of the same background noise region, and μ_r and σ_r^2 are the mean and variance of the R region of interest [37]. Five regions (R = 5) were used in the CNR calculation. The results of these calculations carried out the three test

Table 6.2 Assessment of the proposed denoising algorithm, by means of the SNR and CNR metrics

	SNR		CNR	
	Original	Despeckled	Original	Despeckled
(I1)	9.2	26	2.5	4
(I2)	12.1	31	3	5.9
(I3)	11.5	24	1.9	3.2

images are displayed in Table 6.2. The final perceptual and numerical comparison of original images of I1, I2 and I3 and their despeckled image with proposed method as well as their numerical assessment is given in Fig. 6.11 and Table 6.2 respectively.

It has been also noticed that the median filtered image is more pronounced with respect to contrast. Quantitative assessments taken on the despeckled image showed in Fig. 6.11, shows that the proposed method can provide about 8 dB extra enhancement in SNR and 0.6 dB with respect to CNR, when compared to their counterparts in averaging and median filtering.

Furthermore, the proposed method was seen to exceeded in about 3 dB the SNR in comparison with both, Symmetric Nearest Neighborhood [59] and Wiener noise reduction filters procedures. However, a difference of only 0.2 is observed in CNR.

The performance of the method described here has been confronted with some other existing methods. These results are depicted in Figs. 6.12, 6.13 and 6.14 [53]. Generally speaking a blurring artifact can be depicted in images treated using the method proposed here. On the other hand in one considers Kuwahara [60] image filtering methodologies present a 4 dB SNR and less than 0.1 CNR when compared with simple filtered image. With respect to this issue it is worth to mention here that the fast-real-time effective algorithm described here could be enhanced if one carries out a more accurate sigma estimation, by for example employing an improved version of ANN and using a more accurate mathematical model of noise evaluation. Also, image segmenting can still be improved. These issues are worth to be further analyzed. Finally, due to the generality of the ANN algorithm proposed here, it can be used as a signal processing methodology in other image based techniques, such as photoacoustic imaging systems [61].

6.3 Conclusions

Enhancement of OCT images quality, with respect to speckle reduction methods is of great interest generally for all image based analyzing techniques. In this chapter, the speckle issues in optical coherence tomography images are described, along with some of the developed technologies and algorithms to resolve it. Among software based methods, a speckle reduction algorithm is presented based on the approximation that speckle noise has a Rayleigh distribution represented by a noise parameter, sigma. In the approach analyzed here, a combination of several Multi-Layer Per-

Fig. 6.12 Comparative presentation of six despeckling methods on original OCT test images acquired from the optic nerve region of retina, of a volunteer (AP), white male, provided by Adrian Podoleanu's lab. **a** Original B-scan image of optic nerve, lateral size ~1–1.2 mm, **b** the B-scan image after average filtering (window size: 3), **c** the B-scan image after median filtering (window size: 3), **d**, the B-scan image after wiener filtering (window size: 3), **e** the B-scan image after Kuwahara filtering (window size: 5), **f** the B-scan image after SNN filtering (window size: 3), **g** the B-scan image after using the proposed method [53]

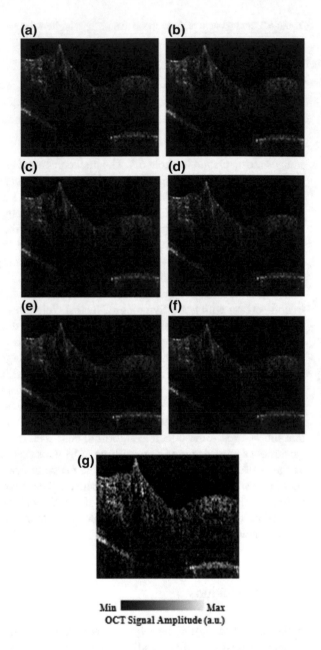

ceptron (MLP) neural networks was used to estimate the noise parameter, sigma, for despeckling retina optical coherence tomography images. The intelligence inherited in the method allows improving the image quality while preserving the edges. The sigma estimator kernel was treated with more than 99.3% reliability on average. Both SNR and CNR values computed for the original and despeckled images

Fig. 6.13 Comparative presentation of six despeckling methods on an original OCT test images acquired from the retina (optic nerve region) of a volunteer (AB-fovea), white male. **a** Original B-scan image of optic nerve, **b** the B-scan image after average filtering (window size: 3), **c** the B-scan image after median filtering (window size: 3), **d** the B-scan image after Wiener filtering (window size: 3), **e** the B-scan image after Kuwahara filtering (window size: 5), **f** the B-scan image after SNN filtering (window size: 3), **g** the B-scan image after using the proposed method [53]

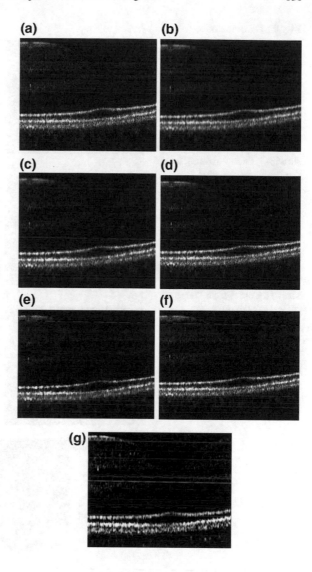

revealed that the method presented here is effective in speckle artifact reducing in the OCT images. Furthermore, the algorithm proposed here is quite satisfactory when with bilateral digital filters commonly used. Finally, due to the generality of proposed ANN algorithm, it can be promptly implemented as a signal processing methodology for treating OCT images from diverse parts of body and as well for different OCT imaging techniques.

Fig. 6.14 Comparative presentation of six despeckling methods on an original OCT test images acquired from the retina (optic nerve region) of a volunteer (AB-fovea), white male. **a** Original B-scan image of optic nerve, **b** the B-scan image after average filtering (window size: 3), **c** the B-scan image after median filtering (window size: 3), **d** the B-scan image after Wiener filtering (window size: 3), **e** the B-scan image after Kuwahara filtering (window size: 5), **f** the B-scan image after SNN filtering (window size: 3), **g** the B-scan image after using the proposed method. The vertical axis is z-axis [53]

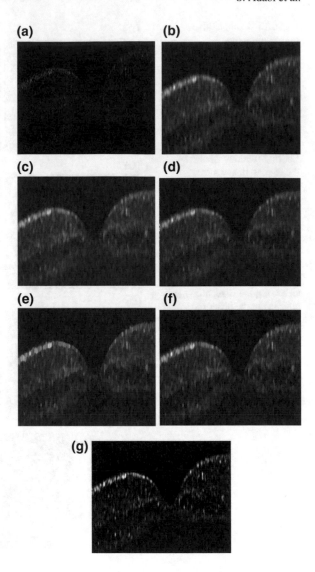

References

1. R.G. Cucu, A.G. Podoleanu, J.A. Rogers, J. Pedro, R.B. Rosen, Combined confocal/en face T-scan-based ultrahigh-resolution optical coherence tomography in vivo retinal imaging. Opt. Lett. **31**, 1684–1686 (2006)
2. E. Hecht, A. Zajac, Optics (1974)
3. A.F. Fercher, W. Drexler, C.K. Hitzenberger, T. Lasser, Optical coherence tomography-principles and applications. Rep. Prog. Phys. **66**, 239 (2003)
4. J.M. Schmitt, Optical coherence tomography (OCT): a review. IEEE J. Sel. Top. Quantum Electron. **5**, 1205–1215 (1999)
5. A.G. Podoleanu, Optical coherence tomography. Br. J. Radiol. (2014)

6. M. Hughes, High lateral resolution imaging with dynamic focus. Ph.D. thesis, University of Kent (2010)
7. V.V. Tuchin, Coherent-Domain Optical Methods: Biomedical Diagnostics, Environment and Material Science (Springer Science & Business Media, 2004)
8. B.K. Alsberg, A.M. Woodward, M.K. Winson, J. Rowland, D.B. Kell, Wavelet denoising of infrared spectra. Analyst **122**, 645–652 (1997)
9. R. Leitgeb, C. Hitzenberger, A. Fercher, Performance of fourier domain vs. time domain optical coherence tomography. Opt. Express **11**, 889–894 (2003)
10. M.R. Avanaki, A. Hojjat, A.G. Podoleanu, Investigation of computer-based skin cancer detection using optical coherence tomography. J. Mod. Opt. **56**, 1536–1544 (2009)
11. M.R. Avanaki, A. Hojjatoleslami, M. Sira, J.B. Schofield, C. Jones, A.G. Podoleanu, Investigation of basal cell carcinoma using dynamic focus optical coherence tomography. Appl. Opt. **52**, 2116–2124 (2013)
12. J. Welzel, Optical coherence tomography in dermatology: a review. Skin Res. Technol. **7**, 1–9 (2001)
13. L.L. Otis, M.J. Everett, U.S. Sathyam, B.W. Colston, optical coherence tomography: a new imaging: technology for dentistry. J. Am. Dent. Assoc. **131**, 511–514 (2000)
14. G.J. Tearney, M.E. Brezinski, B.E. Bouma, S.A. Boppart, C. Pitris, J.F. Southern, J.G. Fujimoto, In vivo endoscopic optical biopsy with optical coherence tomography. Science **276**, 2037–2039 (1997)
15. I.-K. Jang, B.E. Bouma, D.-H. Kang, S.-J. Park, S.-W. Park, K.-B. Seung, K.-B. Choi, M. Shishkov, K. Schlendorf, E. Pomerantsev, Visualization of coronary atherosclerotic plaques in patients using optical coherence tomography: comparison with intravascular ultrasound. J. Am. Coll. Cardiol. **39**, 604–609 (2002)
16. W. Drexler, J.G. Fujimoto, Optical coherence tomography in ophthalmology. J. Biomed. Opt. **12**, 041201-041201-2 (2007)
17. J.G. Fujimoto, C. Pitris, S.A. Boppart, M.E. Brezinski, Optical Coherence Tomography: an emerging technology for biomedical imaging and optical biopsy. Neoplasia **2**, 9–25 (2000)
18. A.M. Zysk, F.T. Nguyen, A.L. Oldenburg, D.L. Marks, S.A. Boppart, Optical coherence tomography: a review of clinical development from bench to bedside. J. Biomed. Opt. **12**, 051403-051403-21 (2007)
19. W. Drexler, J.G. Fujimoto, *Optical Coherence Tomography: Technology and Applications* (Springer Science & Business Media, 2008)
20. J.M. Schmitt, A. Knuttel, M. Yadlowsky, M. Eckhaus, Optical-coherence tomography of a dense tissue: statistics of attenuation and backscattering. Phys. Med. Biol. **39**, 1705 (1994)
21. L. Thrane, M.H. Frosz, T.M. Jørgensen, A. Tycho, H.T. Yura, P.E. Andersen, Extraction of optical scattering parameters and attenuation compensation in optical coherence tomography images of multilayered tissue structures. Opt. Lett. **29**, 1641–1643 (2004)
22. J.M. Schmitt, S. Xiang, K.M. Yung, Speckle in optical coherence tomography. J. Biomed. Opt. **4**, 95–105 (1999)
23. M.R. Avanaki, A.G. Podoleanu, J.B. Schofield, C. Jones, M. Sira, Y. Liu, A. Hojjat, Quantitative evaluation of scattering in optical coherence tomography skin images using the extended Huygens-Fresnel theorem. Appl. Opt. **52**, 1574–1580 (2013)
24. A. Hojjatoleslami, M.R. Avanaki, OCT skin image enhancement through attenuation compensation. Appl. Opt. **51**, 4927–4935 (2012)
25. J.W. Goodman, *Speckle Phenomena in Optics: Theory and Applications* (Roberts and Company Publishers, 2007)
26. M.-R. Nasiri-avanaki, S. Hojjatoleslami, H. Paun, S. Tuohy, A. Meadway, G. Dobre, A. Podoleanu, Optical coherence tomography system optimization using simulated annealing algorithm, in *Proceedings of Mathematical Methods and Applied Computing, (WSEAS, 2009)* (2009), pp. 669–674
27. R.F. Wagner, S.W. Smith, J.M. Sandrik, H. Lopez, Statistics of speckle in ultrasound B-scans. IEEE Trans. Sonics Ultrason. **30**, 156–163 (1983)

28. S. Hojjatoleslami, M. Avanaki, A.G. Podoleanu, Image quality improvement in optical coherence tomography using Lucy-Richardson deconvolution algorithm. Appl. Opt. **52**, 5663–5670 (2013)
29. D.C. Adler, T.H. Ko, J.G. Fujimoto, Speckle reduction in optical coherence tomography images by use of a spatially adaptive wavelet filter. Opt. Lett. **29**, 2878–2880 (2004)
30. M.R. Hee, J.A. Izatt, E.A. Swanson, D. Huang, J.S. Schuman, C.P. Lin, C.A. Puliafito, J.G. Fujimoto, Optical coherence tomography of the human retina. Arch. Ophthalmol. **113**, 325–332 (1995)
31. N. Iftimia, B.E. Bouma, G.J. Tearney, Speckle reduction in optical coherence tomography by "path length encoded" angular compounding. J. Biomed. Opt. **8**, 260–263 (2003)
32. T.M. Jørgensen, L. Thrane, M. Mogensen, F. Pedersen, P.E. Andersen, Speckle reduction in optical coherence tomography images of human skin by a spatial diversity method, in *European Conference on Biomedical Optics* (Optical Society of America, 2007), 6627_22
33. P.A. Magnin, O.T. Von Ramm, F.L. Thurstone, Frequency compounding for speckle contrast reduction in phased array images. Ultrason. Imaging **4**, 267–281 (1982)
34. H.L. Resnikoff, O. Raymond Jr., *Wavelet Analysis: The Scalable Structure of Information* (Springer Science & Business Media, 2012)
35. D.J. Smithies, T. Lindmo, Z. Chen, J.S. Nelson, T.E. Milner, Signal attenuation and localization in optical coherence tomography studied by Monte Carlo simulation. Phys. Med. Biol. **43**, 3025 (1998)
36. R.K. Wang, Reduction of speckle noise for optical coherence tomography by the use of nonlinear anisotropic diffusion, in *Biomedical Optics 2005* (International Society for Optics and Photonics, 2005), pp. 380–385
37. P.M. Shankar, Speckle reduction in ultrasound B-scans using weighted averaging in spatial compounding. IEEE Trans. Ultrason. Ferroelectr. Freq. Control **33**, 754–758 (1986)
38. J.W. Goodman, Some fundamental properties of speckle*. J. Opt. Soc. Am. A **66**, 1145–1150 (1976)
39. M. Pircher, E. Go, R. Leitgeb, A.F. Fercher, C.K. Hitzenberger, Speckle reduction in optical coherence tomography by frequency compounding. J. Biomed. Opt. **8**, 565–569 (2003)
40. M.R. Avanaki, R. Cernat, P.J. Tadrous, T. Tatla, A.G. Podoleanu, S.A. HojjatoleslamI, Spatial compounding algorithm for speckle reduction of dynamic focus OCT images. IEEE Photonics Technol. Lett. **25**, 1439–1442 (2013)
41. A. Ozcan, A. Bilenca, A.E. DesjardinS, B.E. Bouma, G.J. Tearney, Speckle reduction in optical coherence tomography images using digital filtering. J. Opt. Soc. Am. A **24**, 1901–1910 (2007)
42. J. Rogowska, M.E. Brezinski, Evaluation of the adaptive speckle suppression filter for coronary optical coherence tomography imaging. IEEE Trans. Med. Imaging **19**, 1261–1266 (2000)
43. P. Puvanathasan, K. Bizheva, Speckle noise reduction algorithm for optical coherence tomography based on interval type II fuzzy set. Opt. Express **15**, 15747–15758 (2007)
44. B. Sander, M. Larsen, L. Thrane, J. Hougaard, T. Jørgensen, Enhanced optical coherence tomography imaging by multiple scan averaging. Br. J. Ophthalmol. **89**, 207–212 (2005)
45. G. Karasakal, I. Erer, Speckle noise reduction in SAR imaging using lattice filters based subband decomposition, in *IEEE International Geoscience and Remote Sensing Symposium, 2007. IGARSS 2007* (IEEE, 2007), pp. 1476–1480
46. J. Kim, D.T. Miller, E. Kim, S. Oh, J. Oh, T.E. Milner, Optical coherence tomography speckle reduction by a partially spatially coherent source. J. Biomed. Opt. **10**, 064034-064034-9 (2005)
47. M. Avanaki, P. Laissue, A.G. Podoleanu, A. Aber, S. Hojjatoleslami, Evaluation of wavelet mother functions for speckle noise suppression in OCT images. Int. J. Graph. Bioinf. Med. Eng. **11**, 1–5 (2011)
48. M. Avanaki, P.P. Laissue, T.J. Eom, A.G. Podoleanu, A. Hojjatoleslami, Speckle reduction using an artificial neural network algorithm. Appl. Opt. **52**, 5050–5057 (2013)
49. M.R. Avanaki, P.P. Laissue, A.G. Podoleanu, A. Hojjat, Denoising based on noise parameter estimation in speckled OCT images using neural network, in *1st Canterbury Workshop and School in Optical Coherence Tomography and Adaptive Optics* (International Society for Optics and Photonics, 2008), 71390E-71390E-9

50. M.R. Avanaki, M.J. Marques, A. Bradu, A. Hojjatoleslami, A.G. Podoleanu, A new algorithm for speckle reduction of optical coherence tomography images, in *SPIE BiOS* (International Society for Optics and Photonics, 2014), 893437-893437-9

51. J.-N. Hwang, S.-R. LAY, A. Lippman, Nonparametric multivariate density estimation: a comparative study. IEEE Trans. Signal Process. **42**, 2795–2810 (1994)

52. H. Park, R. Miyazaki, T. Nishimura, Y. Tamaki, The speckle noise reduction and the boundary enhancement on medical ultrasound images using the cellular neural networks. 電気学会論文誌. C, **127**, 1726–1731 (2007)

53. S. Adabi, S. Conforto, A. Clayton, A.G. Podoleanu, A. Hojjat, M. Avanaki, An intelligent speckle reduction algorithm for optical coherence tomography images, in *Proceeding of the 4th International Conference on Photonics, Optics and Laser Technology, PHOTOPTICS 2016* (SciTePress, 2016), pp. 40–45

54. M. Bashkansky, J. Reintjes, Statistics and reduction of speckle in optical coherence tomography. Opt. Lett. **25**, 545–547 (2000)

55. K. Hornik, M. Stinchcombe, H. White, Multilayer feedforward networks are universal approximators. IEEE Trans. Neural Netw. **2**, 359–366 (1989)

56. M. Minsky, S. Papert, Perceptrons (1969)

57. T.B. Fitzpatrick, Soleil et peau. J Med. Esthet **2**, 33–34 (1975)

58. M.R. Avanaki, A.G. Podoleanu, M.C. Price, S.A. Corr, S. Hojjatoleslami, Two applications of solid phantoms in performance assessment of optical coherence tomography systems. Appl. Opt. **52**, 7054–7061 (2013)

59. D. Harwood, M. Subbarao, H. Hakalahti, L.S. Davis, A new class of edge-preserving smoothing filters. Pattern Recogn. Lett. **6**, 155–162 (1987)

60. M. Kuwahara, K. Hachimura, S. Eiho, M. Kinoshita, Processing of RI-angiocardiographic images, in *Digital Processing of Biomedical Images* (Springer, 1976)

61. M. Nasiriavanaki, J. Xia, H. Wan, A.Q. Bauer, J.P. Culver, L.V. Wang, High-resolution photoacoustic tomography of resting-state functional connectivity in the mouse brain. Proc. Natl. Acad. Sci. **111**, 21–26 (2014)

Chapter 7
Characterization of the Performance of a 7-Mirror Segmented Reflecting Telescope via the Angular Spectrum Method

Mary Angelie Alagao, Mary Ann Go, Maricor Soriano and Giovanni Tapang

Abstract A segmented reflecting telescope made of seven 76 mm concave mirrors, each with a focal length of 300 mm, was characterized. Its performance was evaluated by computing the point spread function (PSF) and comparing it to an equivalent monolithic mirror. Aberrations were added and corrected using a phase retrieval technique called the Gerchberg-Saxton (GS) algorithm to obtain the correction phase that serves as the input to the spatial light modulator (SLM). Results revealed an improvement in the telescope angular resolution as a result of the implemented phase correction. It was also shown that the PSF varies depending on the orientation and number of mirrors added.

M. A. Alagao (✉) · M. A. Go · M. Soriano · G. Tapang
National Institute of Physics, University of the Philippines, Diliman,
Quezon City, Philippines
e-mail: angelie@narit.or.th; anngelie.alagao@gmail.com

M. A. Go
e-mail: m.go@imperial.ac.uk

M. Soriano
e-mail: msoriano@nip.upd.edu.ph

G. Tapang
e-mail: gtapang@nip.upd.edu.ph

M. A. Alagao
National Astronomical Research Institute of Thailand, Chiang Mai, Thailand

M. A. Go
Department of Biomedical Engineering, Imperial College London,
South Kensington Campus, London SW7 2AZ, UK

© Springer Nature Switzerland AG 2018
P. A. Ribeiro and M. Raposo (eds.), *Optics, Photonics and Laser Technology*, Springer
Series in Optical Sciences 218, https://doi.org/10.1007/978-3-319-98548-0_7

7.1 Introduction

The need for deep space exploration has been driving the construction of novel meter-wide ground-based telescopes. The detection of faint astronomical objects requires adequate amount of light to enter the telescope assembly, for which the aperture has to increase allowing more light power to be gathered and also improving the instrument angular resolution, Θ, given by

$$\Theta = \frac{\lambda}{D} \tag{7.1}$$

where λ is the wavelength and D is the diameter of the aperture. To probe deeper in space, a telescope must have an aperture large enough to collect light from faint celestial objects. The relationship between the aperture size and the limiting magnitude of the celestial object is given by [1]

$$M_L \simeq 3.7 + 2.5 \log_{10}(D^2). \tag{7.2}$$

In Fig. 7.1, the limiting magnitude, M_L, is the maximum brightness of the celestial object that can be detected by a telescope with a diameter D [1]. This means that to be able to image celestial objects of brightness, M_L, the telescope must have the right aperture size.

However, there is a technological limit in the construction of large single mirror telescopes (typically around 10 m) primarily due to costs and transportation issues [2]. Segmented mirror design is normally addressed to circumvent these major drawbacks [2]. The general idea of this design is to use small mirrors placed together to act as a single large mirror. Each mirror has a specific shape and a control system for precise position to reduce, if not eliminate, the optical path difference between segments. In spite of the added complexity of this solution, it is less expensive when compared with the construction of a large mirror, and makes easier the tuning of the telescope

Fig. 7.1 The limiting magnitude increases as the aperture diameter is increased

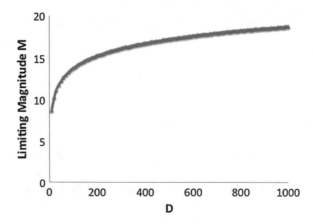

aperture diameter to an optimized value. As a telescope is inevitably a diffraction-limited device, its imaging quality can be assessed through the point spread function (PSF). This function corresponds to the diffraction image of a point through an assumed telescope circular opening. Basically, this is the Fourier transform of the aperture at the exit pupil.

The PSF transverse size is the image of a point in the image plane. The diameter of the first PSF dark ring, the so-called Airy disk, gives a measure of the resolution, with its size determining the lowest feature that the telescope can detect [3]. Actually, the effective PSF detected by a camera is likely to also carry aberrations arising from device misalignment. Corrections can be made to the wavefront by knowing the phase from the difference between the aberrated and ideal.

Aberrations due to atmospheric turbulence and imperfect alignment can be solved by embedding an adaptive optics component in the telescope. An adaptive optics system measures and corrects the incoming wavefront in real time. It is usually composed of the wavefront sensor and the wavefront corrector. Wavefront sensors (WFS), such as the pyramid WFS, Shack-Hartmann WFS and the curvature WFS, measure the shape of the wavefront with sufficient accuracy. On the other hand, wavefront correctors, such as the deformable mirrors and micro-optical-electrical-mechanic systems (MOEMS), compensate the aberrations in the system [4].

A spatial light modulator (SLM) can be used to measure and correct optical aberrations [5], generate Zernike polynomials for aberration simulation [6–8], and perform linear filtering for phase retrieval [9, 10]. The SLM capability for wavefront shaping has been successfully attained in microgear patterns photopolymerization by encoding in the SLM the topological charge and phase level of an optical vortex [11]. In another example, the SLM was used for three-dimensional light reshaping in [12]. This suggests that it is also possible to correct the intrinsic telescope aberrations.

This chapter provides numerical simulations to describe a segmented reflecting telescope composed of seven identical concave mirrors. The aberrations are intentionally added and an SLM is used to shape the wavefront to improve the PSF. We also observed the effect on the PSF as the orientation and the number of mirrors are varied.

7.2   Methods and Procedures

For the segmented reflecting telescope, seven identical mirrors, 76 mm in diameter and a focal length of 300 mm were used. The mirrors are rotated and translated such that they focus at one point. The segmented mirror has an effective diameter of 223.22 mm. The wavelength of the source used in the simulations is 632.8 nm.

Light coming from a star enters the telescope and is focused at the location of the secondary mirror by a segmented mirror. The PSF due to the segmented reflector passes through a 4f-lens system ($f_1 = 100$ mm, $f_2 = 50$ mm). At the Fourier plane of the 4f-lens system, the SLM is inserted to correct for the inherent aberrations.

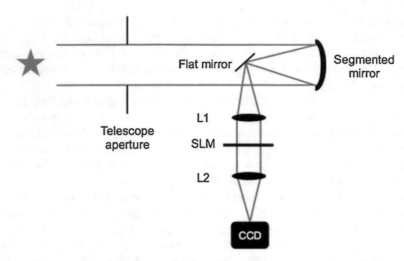

Fig. 7.2 Layout of optical setup used for wavefront correction

The results obtained with the segmented mirror telescope were compared with those attained with monolithic reflecting telescope. For consistency in comparison, both monolithic reflecting and segmented telescopes have the same aperture diameters. The whole system setup is shown in Fig. 7.2.

7.2.1 Calculation of the PSF

The performance of a segmented reflecting telescope was simulated using the angular spectrum method. A plane wave located at $z = 0$ as U(x, y, 0) was considered. Thus, the angular spectrum $\mathcal{U}(fx, fy, 0)$ at this location will be given by the Fourier transform of U(x, y, 0), i.e. [3]

$$\mathcal{U}(f_x, f_y; 0) = \int\limits_{-\infty}^{\infty} \int U(x, y, 0) \exp[-j2\pi\phi]dxdy \qquad (7.3)$$

where

$$\phi = f_X x + f_Y y. \qquad (7.4)$$

The inverse Fourier transform will then give the field amplitude U(x, y, 0), that is,

$$U(x, y, 0) = \mathcal{F}^{-1}\{\mathcal{U}(f_x, f_y; 0)\}. \qquad (7.5)$$

Similarly, the angular spectrum at a z distance will be given by

$$\mathcal{U}(f_X, f_Y; z) = \int\limits_{-\infty}^{\infty} \int U(x, y, z)\exp[-j2\pi\phi)]dxdy \qquad (7.6)$$

while $U(x, y, z)$ the amplitude at a distance z will be denoted as

$$U(x, y, z) = \mathcal{F}^{-1}\{\mathcal{U}(f_x, f_y; z)\}. \qquad (7.7)$$

From (7.3) and (7.6), the relation between $\mathcal{U}(x, y, 0)$ and $\mathcal{U}(x, y, z)$ can be found as

$$\mathcal{U}(x, y, z) = \mathcal{U}(x, y, 0)\exp(j2\pi\sqrt{1 - f_X^2 - f_Y^2}z). \qquad (7.8)$$

Thus, the PSF at point (x, y, z) can be obtained from the angular spectrum method.

7.2.2 Angular Spectrum of the Segmented Mirrors

In Fig. 7.3, the mirrors are represented as apertures.

In order to go with the PSF simulation under this arrangement, the remaining mirrors are rotated about the x and y axis with respect to Mirror 1, which is placed at the origin.

Fig. 7.3 Top view of the seven mirror segments configuration: D is the mirror segment diameter which is set to be 76 mm and θ is the tilt angle

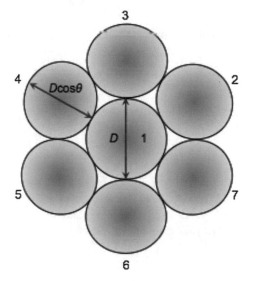

Fig. 7.4 Geometry for
determining the tilt angle of
mirrors

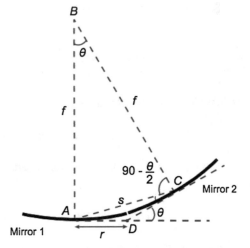

Each mirror has be tilted such that the focus of each mirror overlap. The geometry
for tilt angle calculation is sketched in Fig. 7.4.

The sine law is used in △ABC in order to obtain the relation

$$\frac{s}{\sin \theta} = \frac{f}{\sin(90 - \frac{\theta}{2})}. \tag{7.9}$$

By taking into account △CBD, one can rewrite (7.9) as

$$\frac{s}{\sin(180 - \theta)} = \frac{r}{\sin(\frac{\theta}{2})}. \tag{7.10}$$

Replacing the result of (7.10) into (7.9), the following relation can be obtained

$$\frac{f}{\cos(\frac{\theta}{2})} = \frac{r}{\sin(\frac{\theta}{2})} \tag{7.11}$$

thus, the tilt angle can be calculated from:

$$\theta = 2 \tan^{-1}(\frac{r}{f}). \tag{7.12}$$

Equation 7.12 is used for building up the rotation matrices for each mirror. The
angle attained was found to be 14.4°. By setting z as the optical axis, the angle is
used to rotate about the y-axis and then the x-axis with respect to the mirror position.
The rotation matrix given by

$$R_{xy}(\alpha, \beta) = \begin{pmatrix} \cos\beta & \sin\alpha\sin\beta & \cos\alpha\sin\beta \\ 0 & \cos\alpha & -\sin\alpha \\ -\sin\beta & \sin\alpha\cos\beta & \cos\alpha\cos\beta \end{pmatrix} \qquad (7.13)$$

defines the rotation of each mirror. Together with the translation matrix given by,

$$T_{xy} = \begin{pmatrix} 1 & 0 & dx \\ 0 & 0 & dy \\ 0 & 0 & 1 \end{pmatrix} \qquad (7.14)$$

the new axis (x', y', z') of the off-axis mirrors are calculated using,

$$\begin{pmatrix} x' \\ y' \\ z' \end{pmatrix} = T_{xy} R_{xy} \begin{pmatrix} x \\ y \\ z \end{pmatrix}. \qquad (7.15)$$

The α and β angles are both equal to θ. The dx and dy variables correspond to the translations about the x and y axes, respectively. For the diagonal mirrors, Mirror 2, 4, 5 and 7, the dx is $\sqrt{3}r\cos(\theta)$ and dy $r\cos(\theta)$.

One can also calculate the PSF using the vectorial diffraction approach outlined in [13]. According to this approach, the focus of an aplanatic imaging system can be described by studying the electromagnetic field in the focal region. For this system, the following assumptions are given: (1) it follows the Abbe sine condition, producing a stigmatic image, and (2) the linear dimension of the exit pupil and the distance of the image from the exit pupil must be larger than the wavelength [13].

Figure 7.5 shows how the PSF is computed using the vectorial diffraction approach. Using θ, the relationship between the focal plane of M1 and M2 can be computed. This is the projection of M2 at the focal plane of M1. The projection of M5 is just the reverse of the projection of M2.

Fig. 7.5 The PSF for the other mirrors is computed at the focal plane of M1

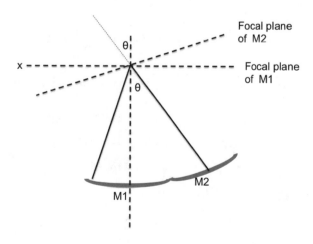

Fig. 7.6 Once the PSF for
M2 and M5 are obtained, it
should be rotated by 60° to
get the PSF for M4 and M7

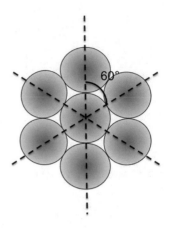

The calculations made for M2 and M5 are also done to compute for the projections
of the pairs M3 and M6, and M4 and M7. However, the PSF computed for M2 and
M5 are added and rotated at 60° to compute for the PSFs due to M4 and M7. This
transverse PSF is flipped to obtain the PSFs due to M3 and M6. This is illustrated in
Fig. 7.6.

7.2.3 Simulations of Aberrations

Aberrations give rise to wavefront deviation which will reduce both the image intensity and contrast. They can be adequately described through Zernike polynomials,
known to be orthogonal over a unit circle.

Using the Zernike polynomials formalism, the aberration function, $W(\rho, \theta)$, can
be expressed in terms of polynomials as follows [14],

$$W(\rho, \theta) = \sum_{j=1}^{\infty} a_j Z_j(\rho, \theta) \tag{7.16}$$

where j is the polynomial ordering number and a_j is the aberration coefficient,

$$Z_{evenj}(\rho, \theta) = \sqrt{(2n+1)} R_n^m(\rho) \cos m\theta \qquad m \neq 0 \tag{7.17a}$$

$$Z_{oddj}(\rho, \theta) = \sqrt{(2n+1)} R_n^m(\rho) \sin m\theta \qquad m \neq 0 \tag{7.17b}$$

$$Z_j(\rho, \theta) = \sqrt{(n+1)} R_n^0(\rho) \qquad m = 0 \tag{7.17c}$$

and

$$R_n^m(\rho) = \sum_{s=0}^{(n-m)/2} \frac{(-1)^s (n-s)!}{s!(\frac{n+m}{2} - s)!(\frac{n-m}{2} - s)!}. \tag{7.18}$$

The following aberrations have been used: coma, astigmatism, spherical aberration and defocus. Aberrations were simulated by means of the following expression

$$PSF_{aberrated} = \mathcal{F}\{\mathcal{F}^{-1}\{PSF_{segmented}\}\exp(i\frac{2\pi}{\lambda}\Phi)\}\}$$ (7.19)

where Φ is given by the following Zernike polynomials [15]:

$$\Phi_{sphericalaberration} = 5\lambda\rho^4$$ (7.20)
$$\Phi_{coma} = 0.3\lambda\rho^3\cos\theta$$ (7.21)
$$\Phi_{astigmatism} = 10\lambda\rho^2\cos 2\theta$$ (7.22)
$$\Phi_{defocus} = 8\sqrt{(3)}(2\rho^2 - 1).$$ (7.23)

7.2.4 Wavefront Correction

For wavefront correction, the correcting phase that will serve as the input to the SLM has to be calculated. The Gerchberg-Saxton (GS) algorithm was used to get the phase information from intensity image. The amplitudes, U(x, y), of the input data can be obtained from PSF intensity image by taking the square root of the measured intensities [3], that is,

$$U(x, y) = \sqrt{I(x, y)}\exp(i\phi(x, y)).$$ (7.24)

From (7.24), a guess phase $\phi(x, y)$ is first taken. This phase is then multiplied by the amplitude of the source image and the Fourier transform is computed. The phase of the obtained complex field was then multiplied with the target image amplitude. The inverse Fourier transform was then applied and checked if the target image has already been reconstructed. This procedure was repeated until a satisfactory reconstruction error is attained [16]. A schematic of this algorithm is shown in Fig. 7.7.

In the simulations carried out here, the aberrated PSF of the segmented mirror was used as the source and the unaberrated PSF of the segmented mirrors was used as the target. The phase information obtained from the GS algorithm was used as the correction phase and the correction success was evaluated by computing $PSF_{corrected}$ from

$$PSF_{corrected} = \mathcal{F}\{|\mathcal{F}^{-1}\{PSF_{aberrated}\}|\exp(i\Phi_{correction})\}.$$ (7.25)

where $\Phi_{correction}$ is the correction phase.

To evaluate the reconstruction quality, the Linfoot's criteria of merit was used, and the following parameters are calculated as:

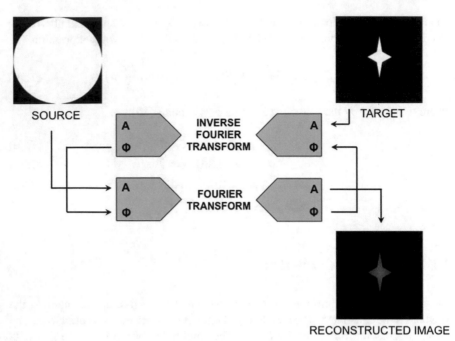

Fig. 7.7 Schematic of the GS algorithm

$$F = 1 - \frac{< (I_{target} - I_{reconstructed})^2 >}{< (I_{target})^2 >} \qquad (7.26)$$

$$Q = \frac{< (I_{reconstructed})^2 >}{< (I_{target})^2 >} \qquad (7.27)$$

$$C = \frac{< |I_{reconstructed}||I_{target}| >}{< (I_{reconstructed})^2 >}. \qquad (7.28)$$

Basically, the criteria states that the signal is perfectly recovered when all of these parameters values are equal to unity. Otherwise, if for example, $C < 1$, the reconstructed profile is narrower than the target profile and if $Q < 1$, both reconstructed and target profiles are erroneously close to each other [17].

7.3 Results and Discussion

The transverse PSFs of a single mirror segment, the segmented mirrors and the monolithic mirror are shown in Fig. 7.8. It can be clearly seen that by adding the 7 mirrors fields, the attained PSF is much smaller than the one for the individual

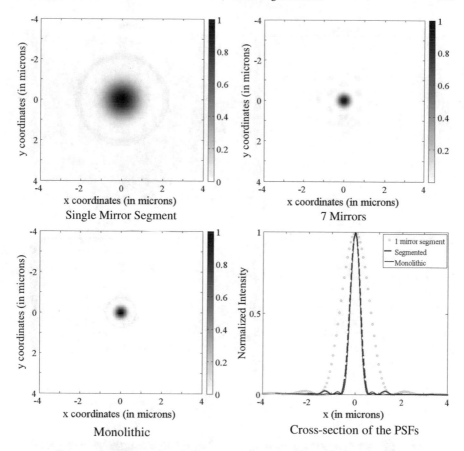

Fig. 7.8 PSFs results obtained for single mirror segment, segmented mirror and monolithic mirror

mirror segment. This result is consistent as the segmented mirrors effective aperture diameter is now larger than that of the single mirror.

By taking the PSF cross-section, it can be seen that the Airy disk width of the segmented mirror is broader than that of the monolithic mirror, although they have the same aperture diameter. This is a direct result from segmentation of the reflecting telescope. Results match calculations done using Richards and Wolf integral [13]. However, this is slower compared to the angular spectrum method.

Figures 7.9 and 7.10 present the results for the aberration correction. By applying the GS algorithm, the correction phase was retrieved and served as the input to the SLM for aberration correction. The quality of the correction was assessed from the obtained Linfoot's merit criteria values for both the aberrated PSF and the corrected PSF. These values are displayed in Table 7.1.

The SLM shown in Fig. 7.2 is a transmitting twisted nematic spatial light modulator such as the Holoeye LC2012. Phase modulation takes place when an electric field

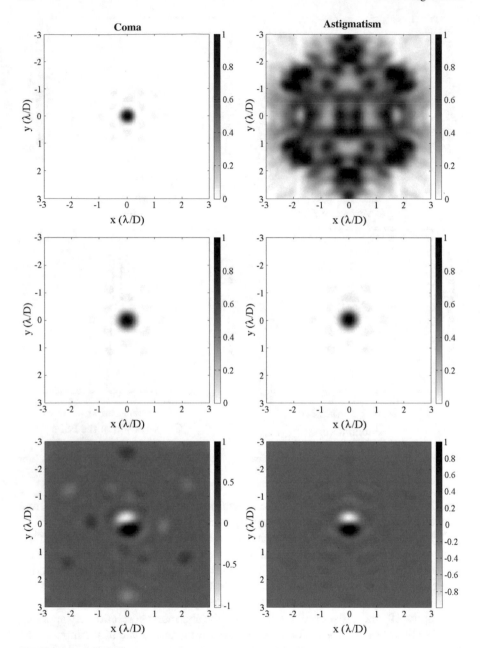

Fig. 7.9 Aberrated (1st row), corrected (2nd row) and the difference between the target and the corrected PSF of the segmented reflecting telescope (3rd row) where λ is equal to 632.8 nm and D is 223.22 mm, the effective diameter of the telescope

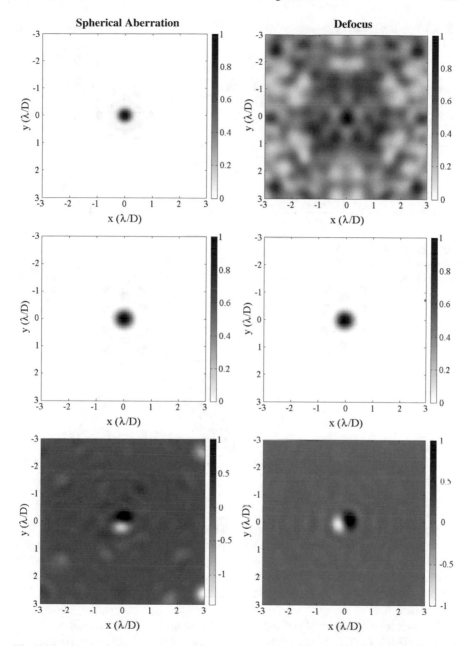

Fig. 7.10 Aberrated (1st row), corrected (2nd row) and the difference between the target and the corrected PSF of the segmented reflecting telescope (3rd row) where λ is equal to 632.8 nm and D is 223.22 mm, the effective diameter of the telescope

Table 7.1 Figure of merit values according to the Linfoot's criteria for both aberrated and corrected PSF, using the angular spectrum method

Aberration type		Fidelity	Structural content	Correlation
Spherical aberration	Aberrated	0.9670	0.7005	0.8337
	Corrected	1.0000	1.0000	0.9988
Coma	Aberrated	0.9956	0.9990	0.9973
	Corrected	1.0000	1.0000	1.0000
Astigmatism	Aberrated	0.0179	0.0137	0.0158
	Corrected	0.9843	1.0000	0.9922
Defocus	Aberrated	0.0045	0.0022	0.0033
	Corrected	0.9843	1.0000	0.9922

Fig. 7.11 The PSF intensity increases as the number of mirrors is increased

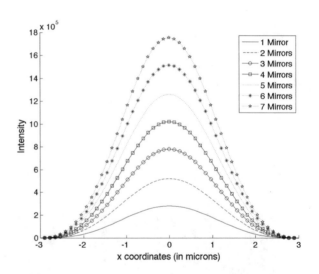

is applied as a result of molecular alignment along electric field direction [18]. The correcting phase, obtained from the GS algorithm, will be sent to the SLM through a video signal from the computer. In the experiment considered here, the PSF should be scaled so that $\triangle f = \frac{\lambda f M}{D}$ where λ is the wavelength, f is the focal length, M is the pixel dimension and D is the aperture diameter. The $\triangle f$ value was computed to be 640 μm.

The effect on the intensity of the PSF as the number of mirrors is increased as shown in Fig. 7.11. This implies that increasing the number of mirrors also increases the light gathering power.

From Figs. 7.12 and 7.13, it can also be seen that the PSF narrows in the direction where the mirrors are added and the extent decreases as the number of mirrors increases. The orientation of mirrors are also varied and its effect on the PSF are

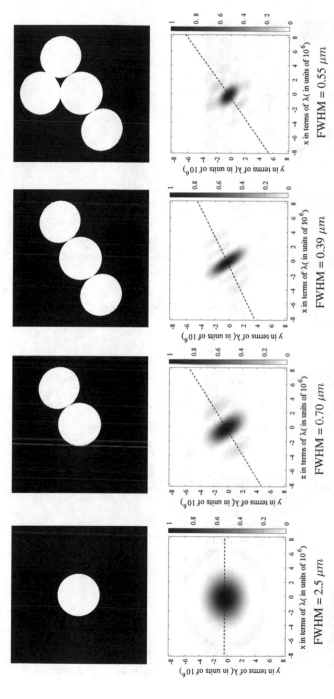

Fig. 7.12 Transverse PSF as the number of mirrors is increased

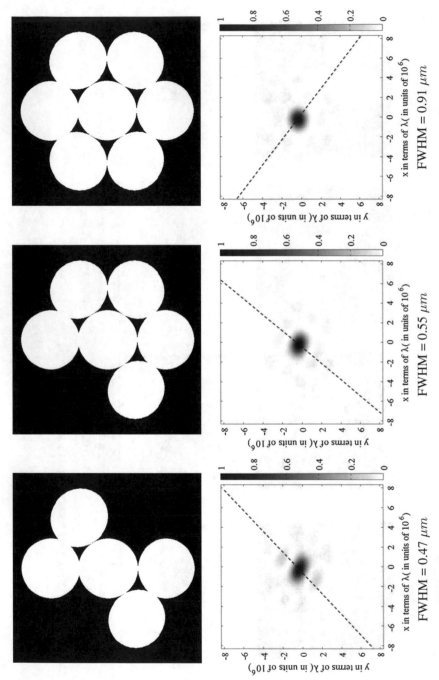

Fig. 7.13 Transverse PSF as the number of mirrors is increased

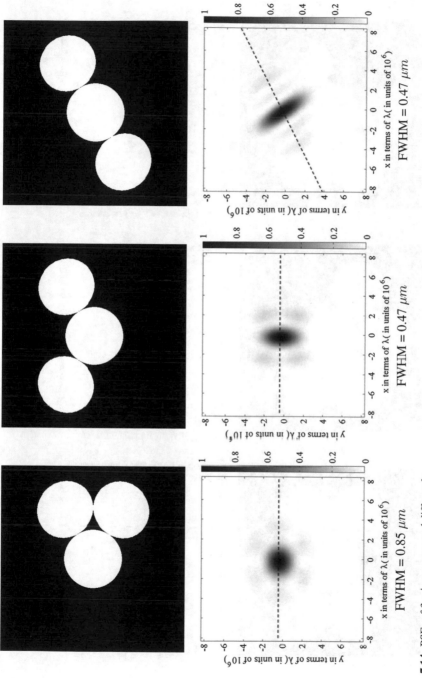

Fig. 7.14 PSFs of 3 mirrors arranged differently

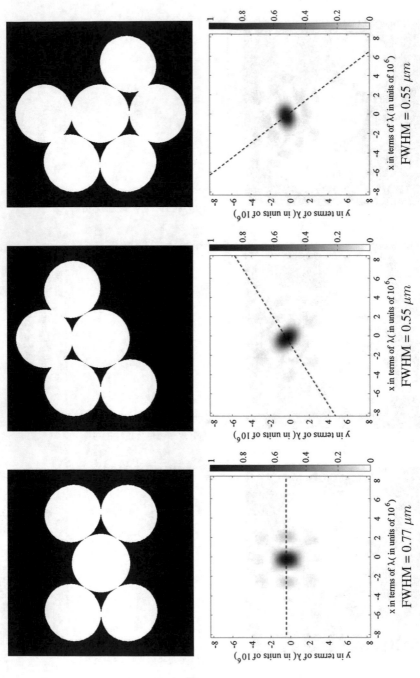

Fig. 7.15 PSFs of 5 and 6 mirrors arranged differently

Fig. 7.16 Corresponding diameter of the monolithic mirror for a given orientation and number of segmented mirrors

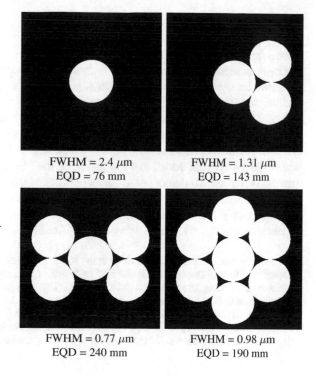

FWHM = 2.4 μm
EQD = 76 mm

FWHM = 1.31 μm
EQD = 143 mm

FWHM = 0.77 μm
EQD = 240 mm

FWHM = 0.98 μm
EQD = 190 mm

shown in Figs. 7.14 and 7.15. The dotted lines indicate the dimension at which the full width at half maximum (FWHM) was calculated.

The equivalent diameter of the monolithic mirror that gives the same FWHM, given the number and orientation of the segments, were also calculated. Based on Fig. 7.16, it can be seen that the equivalent diameter of a monolithic mirror that has the same FWHM as the segmented mirror is only 190 mm, less than the actual effective diameter which is 223.22 mm.

It should be worth mentioning that the use of identical mirrors reduces both the cost and complexity in constructing very large aperture telescopes. It does not require the need for precisely shaping the mirrors such that each mirror segment must be accurately polished to the edge in order to perfectly form a large mirror. Assembling the mirror will also be effortless as the structural support will not be dependent on the mirror segments individual shape. In addition, placing the SLM adaptive optics device before the camera permits the incoming wavefront to be changed in order to compensate for any misalignment or aberrations inherent to the telescope system.

7.4 Conclusion

A technique for improving the angular resolution of a segmented reflecting telescope has been numerically evaluated. The technique was applied to seven identical concave mirrors configured such that their foci coincide, removing the requirement of precise shaping of the individual mirror segments. It was shown that the segmentation of the mirror caused the PSF to broaden. This resulted to an FWHM that is equivalent to a monolithic mirror with a smaller diameter. Upon varying the orientation and the number of mirrors, it was observed that the PSF narrows in the direction where the mirrors are added and the extent of the PSF decreases as the number of mirrors increases. To demonstrate the capability of the wavefront correction using the SLM, aberrations were deliberately added. The GS algorithm was used to obtain the correction phase that serves as the SLM input. The wavefront correction success depends on the phase retrieval using the GS algorithm.

Acknowledgements This work was done at the National Institute of Physics, University of the Philippines, Diliman, Quezon City. It was made possible by the support from the DOST PCIEERD Standards and Testing Automated Modular Platform (STAMP) project (03439), the University of the Philippines Office of the Vice President for Academic Affairs EIDR-VISSER::SM project (C02-001) and the UP System Enhanced Creative Work and Research Grant (ECWRG 2014-11). It was also partly funded by the UP OVPAA ECWRG. GT and MAA would like to thank Dr. Paul Leonard Hilario and Dr. Caesar Saloma for their invaluable advice in this work.

References

1. I. Ridpath, *Norton's Star Atlas and Reference Handbook* (Prentice Hall, 1998)
2. G. Chanan, J. Nelson, T. Mast, Segmented mirror telescopes, in *Planets, Stars and Stellar Systems. Volume 1: Telescope and Instrumentation* (Springer, 2013), pp. 99–136
3. J. Goodman, *Introduction to Fourier Optics*, 3rd edn. (Roberts and Compan Publishers, 2005)
4. M. Kasper, R. Davies, Adaptive optics for astronomy. Annu. Rev. Astron. Astrophys. (2012)
5. J. Arines et al., Measurement and compensation of optical aberrations using a spatial light modulator. Opt. Soc. Am. (2007)
6. G. Love, Wavefront correction and production of Zernike modes with a liquid-crystal spatial light modulator. Appl. Opt. **36**, 1517–1524 (1997)
7. K.A. Nugent, T.E. Gureyev, A. Roberts, Phase retrieval with the transport-of-intensity equation: matrix solution with use of Zernike polynomials. J. Opt. Soc. Am. **12**, 1932–1941 (1995)
8. A. Vyas, B. Ropashree, R.K. Banyal, B. Raghavendra, Spatial light modulator for wavefront correction, in *Proceedings of International Conference on Trends in Optics and Photonics* (2009), pp. 318–330
9. J. Astola, V. Katkovnik, Phase retrieval via spatial light modulator phase modulation in 4f optical setup: numerical inverse imaging with sparse regularization for phase and amplitude. J. Opt. Soc. Am. **29**, 105–116 (2012)
10. C. Kopylow, R.B. Bergmann, C. Falldorf, M. Agour, Phase retrieval by means of a spatial light modulator in the Fourier domain of an imaging system. Appl. Opt. **49**, 182601830 (2010)
11. G. Bautista, M.J. Romero, G. Tapang, V.R. Daria, Parallel two-photon photopolymerization of microgear patterns. Opt. Commun. **282**, 3746–3750 (2009)
12. P.L.A. Hilario, M.J. Villangca, G. Tapang, Independent light fields generated using a phase only spatial light modulator. Opt. Lett. **39**, 2036–2039 (2014)

13. E. Wolf, B.R.: Electromagnetic diffraction in optical systems. II. Structure of the image field in the aplanatic system. Proc. R. Soc. Lond. Ser. A Math. Phys. Sci. **253**: 358–379 (1959)
14. V. Mahajan, Zernike circle polynomials and optical aberrations of systems with circular pupils. Suppl. Appl. Opt. **33**, 8121–8124 (1994)
15. M. Born, R. Wolf, *Principle of Optics* (Cambridge University Press, 1999)
16. R.W. Gerchberg, W.O. Saxton, A practical algorithm for the determination of phase from image and diffraction plane pictures. Optik **35**, 237–246 (1972)
17. G. Tapang, C. Saloma, Behavior of the point-spread function in photon-limited confocal microscopy. Appl. Opt. **41**, 1534–1540 (2002)
18. Laser Teaching Center, D.o.P., Astronomy, S.U., *An Introduction to Spatial Light Modulators*

Chapter 8
CMOS Silicon Photomultiplier Development

N. D'Ascenzo, V. Saveliev and Q. Xie

Abstract CMOS—complementary metal-oxide-semiconductor technology at present time is the most advanced semiconductor technology for the development and production of microelectronic elements. Many signs have long pointed toward CMOS as the preferable sensor technology of the future. In many ways, the bright future for CMOS sensor technology was made officially by leading electronic companies in early 2015 to claim that up to 2025 all kind of sensors will be produced in CMOS technology. Beyond this, improvements of CMOS technology and the strong price/performance ratio of CMOS sensors make them increasingly attractive for many academic and industrial applications. The development of Silicon Photomultiplier (SiPM)—a new sensor for the low photon flux in standard CMOS technology—is an important new step for the development, optimisation and mass production of SiPM for the wide application areas as nuclear medicine, experimental physics, visualisation systems. It is an important step for the future developments of new SiPM structures, especially advanced digital SiPM structures, digital SiPM imagers, and Avalanche Pixel structures (APix) for the detection of ionisation particles.

8.1 Introduction

For more than 60 years Photomultiplier Tubes (PMTs) have been filling the area of detection of low photon flux practically without alternative [1], despite the fact that the many critical fundamental problems of this devices are very well known, as example strong influence of magnetic field and others.

Regarding current semiconductor structures for low photon flux detection, only few options have been explored. The main critical problem to develop semiconductor

N. D'Ascenzo · V. Saveliev (✉) · Q. Xie
Huazhong University of Science and Technology, Wuhan, China
e-mail: saveliev@hust.edu.cn; saveliev@mail.desy.de

N. D'Ascenzo
e-mail: ndasc@hust.edu.cn

Q. Xie
e-mail: qgxie@hust.edu.cn

© Springer Nature Switzerland AG 2018
P. A. Ribeiro and M. Raposo (eds.), *Optics, Photonics and Laser Technology*, Springer
Series in Optical Sciences 218, https://doi.org/10.1007/978-3-319-98548-0_8

device for the extremely low photon flux is the relative high level of thermal noise of the semiconductor detector structures and the associated front-end electronics. The solution of this problem can be done in two ways: cooling the photosensor and the related front-end electronics up to cryogenic temperatures or providing a very high intrinsic amplification, up to 10^6, as comparable to the PMT amplification.

A way to overcome the noise issue is Visible Light Photon Counters (VLPC) [2]. These photosensors are semiconductor avalanche structures operated at the temperature of 4 K, in order to suppress the thermal noise. The results were successful and showed the possibility to detect low photon flux up to a single photon. However the operational conditions were too complicated to be acceptable for a wide area application as this sensor needed to be operated a cryostat down to a temperature of 4K. This is essentially a major drawback even under laboratory conditions.

The development of modern low photon flux detection structures, known at present time as Silicon Photomultipliers, was initiated at the beginning of the '90s, starting with the investigation of Silicon Metal Oxide Semiconductor (MOS) structures in avalanche breakdown operation mode towards single photon detection. The results revealed to be promising, however the strong limitation was the requirement of external recharge circuits for the device discharge after charging the MOS structure during the photons detection.

The next issue deals with the development of special resistive layers instead of oxide layers. These so-called Metal Resistive Semiconductor (MRS) structures allow to recharge the device after photon detection and in additionally enables to control the breakdown avalanche process through quenching. Such a structures had very high and stable amplification suitable for single photons detection, in comparison to conventional avalanche photodetector structures, but limited sensitive area.

The idea of Silicon Photomultiplier or more strictly Silicon Photoelectron Multipliers was undertaken to circumvent the major drawbacks of the above mentioned structures, namely small sensitive areas due to the instability of amplification over large area, low dynamic range and slow response.

Under this compliance it was chosen to create semiconductor structures consisting of space distributed fine metal resistor semiconductor micro sensors having individual quenching and common output. The attained results were in fact impressive leading to a high resolution clear detection of single photon spectra on the semiconductor structure at room temperature.

The first proposed Silicon Photomultiplier had fine silicon structure of avalanche breakdown mode micro-cells with a common resistive layer as quenching recharged element and parallel connections of micro-cells to a common output electrode [3].

The next step in the development of the Silicon Photomultiplier was the optimization of the detection structures, namely in increasing the geometrical efficiency. This characteristic is the ratio of the photon-sensitive-area to the total area of the silicon photomultiplier. Its increase allows for getting an higher detection efficiency. Further optimization goals were the tuning of the optimal operation condition in terms of bias and time performance, and also the technological processes improvement.

With the advances in technology in the middle of the '90s, the micro-cells were possible to be placed a closer to each other and the common resistive layer, as quenching element, was replaced by individual integrated resistors coupled to the each micro-cell with clear optimization of both position and size. In this way the actual silicon photomultiplier devices started to be available for the applications [4].

The new problem of the optimized silicon photomultipliers structures was the optical crosstalk in fine detection structure arising from light emission during the avalanche breakdown processes in silicon [5]. In Silicon Photomultipliers with tiny space between the micro-cells, the chance of secondary photons detection by neighbourhood micro-cells is large in such a inevitably has be considered. This problem is mainly affecting the area of very low photon flux detection, where the optical crosstalk is able to dramatically change the measurement results. This problem sorted out through modern technology processes, which physically provided optical isolation of the micro-cells within the integrated structure level. For the optical crosstalk suppression between the micro-cells, a trench structure was built around the micro-cells and filled with a non transparent optic material. The most recent developments achieved in this technology allowed the achievement of high performance in detection of very low photon flux and have originated a special class of silicon photomultipliers, the so-called quantum photo detectors (QPD) [6].

The Silicon Photomultiplier is in fact the first semiconductor device, which could not only compete with traditional photomultiplier tubes, in terms of low photon flux detection, but also presenting great advantages in performance and operation conditions. Thus there is a has great future ahead in many areas of applications such as experimental physics, nuclear medicine, homeland security, military applications and many other. Silicon Photomultipliers show an remarkable performance including the single photon response at room temperature. As a remark intrinsic gain of multiplication can reach 10^6, high detection efficiency can be as high as 25–60% for the visible range of light and time responses of 30 ps can be attained in these devices. In addition, conditions of operation are suitable for many applications as for example: operation bias between 20 and 60 V; operated temperature at room temperature or under cooling conditions; and insensitive to electromagnetic fields. The fabrication of these devices at the light of actual semiconductor technology, thus compatible with mass production, envisages the production of compact devices.

In the same time all developments of the SiPM is done on the basis of a special semiconductor technology, which actually is not necessary. The semiconductor material for the SiPM does not require the special characteristic as high resistivity, or high purity and can use of standard CMOS technology materials. This feature opens the way to use the modern advanced technology as CMOS technology for the development and production of the conventional SiPMs and speed up the progress of the future development of SiPM, creating principal new method of signal processing, in particular direct conversion of the micro-sensor signal to digital form, as the

Digital Silicon Photomultiplier which combine the sensors with the modern processing electronic on the same substrate and generally use the all modern development in microelectronic area [7–10].

8.2 The CMOS Silicon Photomultiplier Technology Design

A detailed description of the Silicon Photomultiplier structure and operational principle can be found in [11–13]. In this study we focus on the implementation and optimization of the structures and performances of Silicon Photomultipliers by using standard CMOS technology [14, 15]. This topic has currently a strong research interest as first step to future developments of the advanced structures of Silicon Photomultipliers.

Few not trivial issues are faced in the design of SiPM in standard CMOS technology, here we are will discuss of two main critical points of SiPM implementation. First, the achievement of the proper breakdown condition corresponding to an electric field configuration localized in particular region, corresponding to the micro-cells sensitive area. Second, the design of quenching elements, highly resistive characteristics and integrated within the micro-cell, guaranteeing small thickness, small size, precision at the level of few %.

The main problem is the form of the uniform high electric field on the all area of micro-cells to satisfy the breakdown conditions of pn-junctions. Critical areas in this case are formed at the sharp boundary of breakdown pn-junctions and particularly in the corners of the breakdown pn-junctions. The method of prevent the localized breakdown conditions at the border of the breakdown pn-junctions is using so called guard ring structures, which changes the conditions of breakdown in this regions. The CMOS technology offers the possibility of guard rings structures implementation for the planar pn-junctions, but it is not trivial [14, 16, 17, 24].

Two Silicon Photomultiplier structures of sensitive micro-cells, will be described here for possibility of implementation within CMOS processes. The basis of the SiPM sensor micro-cell is the planar pn-junction ($p^{++}/nwell$) which optimised for operation in the avalanche breakdown mode. Structures of the two micro-cells shown under analysis shown in Fig. 8.1. For the up structure the guard ring is formed by use the shallow trench isolation (STI) process, which is provided as standard option in CMOS technology. The lower SiPM sensor micro-cell structure has so called a virtual guard ring, formed as special regions on the periphery with different doping concentration. In the second structure STI are formed close to the p^+ implantation as a standard procedure in the CMOS process, but kept "far" from the sensor breakdown micro-cell, with explanation below.

The structures are operated in breakdown mode and a quenching element is required in order to quench of the self-sustaining avalanche process and recovery the pn-junction. This study we uses a passive quenching element consists of a high resistor integrated close to the sensitive micro-cell, for simplicity is not shown on the figures. The high value resistor is connected to p^{++} of the structures. The standard

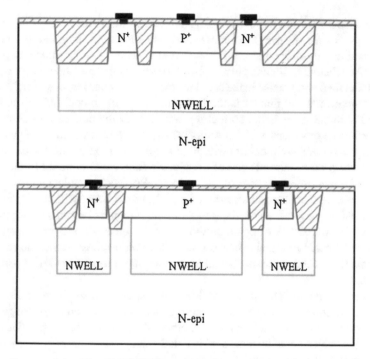

Fig. 8.1 Cross section of the CMOS SiPM avalanche breakdown micro-cell structures, with STI guard ring (upper) and Virtual Guard Ring (down)

CMOS technology processes include High Resistive Polysilicon as a standard option, but it is required special design of the quenching resistor to reach the requirement characteristics. This question will not discuss in scope of this study because for the design of the sensor micro-cells the requirements for the quenching resistors can be relax in particular for geometrical design.

The output is standard common electrode, in case of array of the sensitive avalanche breakdown micro-cells.

8.3   Mathematical Modelling of the CMOS Silicon Photomultiplier

The mathematical modelling of the CMOS Silicon Photomultiplier detection structure plays a central role in the design of the structures in term of electrical performances and optic performances. Following the recent trends in CMOS technology design, in fact, a mathematical model with strong predictive power offers a guidance in the choice of the technology processes minimising the risks and the costs of a less efficient trial-and-error procedure.

With respect to other photodetectors structures fabricated using the standard CMOS processes, whose theoretical understanding and simulation are well-established due to their large use in the manufacturing of electronics components, the CMOS Silicon Photomultiplier required to solve of the specific problems, which is not so detailed and precise in the standard tools of simulations, as example Silvaco TCAD framework. On this basis the key-problem in the modelling of the CMOS Silicon Photomultiplier is the correct reproduction of the conditions of the break-down avalanche processes, which are very sensitive to the correct chose of parameters and time consuming computation with high precision. In particular the calculations of the electron-holes currents and of the electric field corresponding to the applied external bias requires a precision of millivolts range in correspondence to a current of nano-amperes and an internal electric field of about 10^5 V/cm within microns scale. Such precision goes well ahead the standard requirements of electronic components simulation, as transistors or common diodes, and is based on a correct selection of a list of reasonable physical processes as well as on a robust numerical algorithm and as mentioned above the substantial computation time on the powerful computing systems.

The task of the mathematical modelling consist of the estimation of the electric field structure corresponding to the technological processes parameters and geometry of the sensitive micro-cells. Finally the mathematical model should provide a set of observables, which could be measured in a dedicated experimental samples and help the guidance of the design and technological choices.

8.3.1 Physics Processes

The physics processes involved in the avalanche breakdown mode are identify as a typical coupled problem, in which the conservation of the number and energy of electrons and holes should satisfy the additional constrain in the strong electric field generation, when an external bias is applied to the structure.

The relation of electrostatic potential with space charge density is accounted by Poisson's Equation:

$$\nabla \left(\epsilon \nabla \phi \right) = -\rho \tag{8.1}$$

where:

ρ is the local space charge density
ϕ is the electrostatic potential
ϵ is the local permittivity.

The electric field $\mathbf{E}$ is defined from the electrostatic potential as $\mathbf{E} = -\nabla \phi$.

The continuity equations for electrons and holes are ruled by the following equations:

$$\frac{\partial n}{\partial t} = \frac{1}{q} \nabla \cdot \mathbf{J_n} + G_n - R_n \tag{8.2}$$

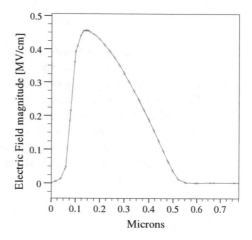

Fig. 8.2 1-dimensional profile of the electric field obtained on a typical CMOS SiPM detection structure

$$\frac{\partial p}{\partial t} = -\frac{1}{q}\nabla \cdot \mathbf{J_p} + G_p - R_p \tag{8.3}$$

where:

n is the electron concentration
p is the holes concentration
$\mathbf{J_n}$ is the electron current density
$\mathbf{J_p}$ is the holes current density
G_n is the generation rate for electrons
G_p is the generation rate for holes
R_n is the recombination rate for electrons
R_p is the recombination rate for holes
q is the electric charge of an electron.

The generation rate $G_{n,p}$ for electrons and holes plays a central role in the correct physical modelling of the SiPM detection structure. When the electric field is strong enough, typically stronger than 10^5 V/cm, free carriers are accelerated, reach an energy higher than the ionization energy for electrons and holes and generate more free carriers in collisions with the atoms of the crystals. Such process is called impact ionization. For the avalanche breakdown process is required extremely high electric field that the impact ionization appears for both type of carriers and free paths are less then the depletion thickness.

The 1-dimensional profile of the electric field corresponding to a considered CMOS SiPM avalanche breakdown structure is shown on Fig. 8.2, calculated using (8.1), is shown on Fig. 8.2. The elective field reaches a maximal value of approximately 4.4×10^5 V/cm within the pn-junction.

The impact ionization process is described by the impact ionization of electrons and holes, respectively α_n and α_p, as:

$$G_{n,p} = \alpha_{n,p} \mathbf{J}_{n,p} \tag{8.4}$$

The ionization coefficient represents the number of electron-hole pairs generated by each carrier per unit distance. The correct evaluation of this parameter and its experimental validation is by far the most important part of the simulation of the CMOS Silicon photomultiplier. The estimation of the breakdown voltage, the working current and the response of the sensitive micro-cell depends on the value of the impact ionization parameter. It is important taking to account the temperature dependent impact ionization, because our goal is develop the sensor micro-cell operated in room temperature, and more precise and realistic estimation of the ionization rate in the sensitive area [18].

The recombination rate $R_{n,p}$ for electrons and holes is calculated according with the Shockley-Read-Hall [19] and three particle transition Auger models [20] and describes the phonon transition in the presence of a trap within the forbidden gap of the semiconductor.

The mathematical model allows to make the detailed analysis for the contribution of the different processes to the total current. On Fig. 8.3 shown the IV-curve, corresponding to a considering CMOS SiPM avalanche breakdown sensor micro-cell structure and the main contributions. The full dots represent the total current. The separated components correspond to the recombination (Auger and Shockley-Read-Hall) and the impact ionization processes. An additional line on the plot corresponds to a physical model without the impact ionization process. Up to approximately 8 V the recombination current from Auger and Shockley-Read-Hall is dominant. At around 8 V the electric field strength is high enough for a significant contribution of the impact ionization process, which becomes clearly dominant at the breakdown. We observe that at breakdown the recombination current has a drop. The curve corresponding to the physical model without impact ionization shows that the recombination current is by far the leading current above the breakdown voltage and determines the dark current level of the CMOS SiPM sensor micro-cell when an

Fig. 8.3 The main current contributions corresponding to the recombination (Auger and Shockley-Read-Hall) and the impact ionisation processes in the mathematical model of a CMOS SiPM sensor avalanche breakdown micro-cell structure

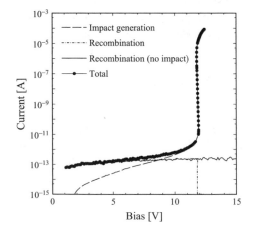

avalanche is not triggered by a photon or a thermal electron-hole pair (should not to be confused with dark rate).

An additional equation is needed for the estimation of the carriers temperature. We use the energy balance transport model consisting of the equations for electrons and holes:

$$\nabla \cdot \mathbf{S}_n = \frac{1}{q}\mathbf{J}_n \cdot \mathbf{E} - W_n - \frac{3k}{2}\frac{\partial}{\partial t}\lambda_n n T_n \tag{8.5}$$

$$\mathbf{J}_n = qD_n\nabla n - q\mu_n n\nabla\psi + qnD_n\nabla T_n \tag{8.6}$$

$$\mathbf{S}_n = -K_n\nabla T_n - \left(\frac{k\delta_n}{q}\right)\mathbf{J}_n T_n \tag{8.7}$$

$$\nabla \cdot \mathbf{S}_p = \frac{1}{q}\mathbf{J}_p \cdot \mathbf{E} - W_p - \frac{3k}{2}\frac{\partial}{\partial t}\lambda_p p T_p \tag{8.8}$$

$$\mathbf{J}_p = qD_p\nabla p - q\mu_p p\nabla\psi + qpD_p\nabla T_p \tag{8.9}$$

$$\mathbf{S}_p = -K_p\nabla T_p - \left(\frac{k\delta_p}{q}\right)\mathbf{J}_p T_p \tag{8.10}$$

where:

S_n is the electron energy flux density
S_p is the holes energy flux density
W_n is the energy density loss for electrons
W_p is the energy density loss for holes
T_n is the temperature of electrons
T_p is the temperature of holes
k is the Boltzmann constant
λ is a constant related to the Fermi or Boltzmann statistics
D_n is the thermal diffusivity for electrons
D_p is the thermal diffusivity for holes
K_n is the thermal conductivity for electrons
K_p is the thermal conductivity for holes.

The electrons and holes thermal diffusivities are coming from the carriers frictional interaction with lattice and also between themselves. As carriers are accelerated by the electric field they loose momentum as a result of scattering processes. These include phonons, carriers and impurity scattering, surface and material imperfections. This effect is accounted with a carrier mobility parameter, which is dependent of the electric field, lattice temperature and doping concentration and defines the carriers thermal diffusivity. In the low electric field regions of the detection structure, the relation between mobility and carriers concentration is introduced in the model by means of a experimental look-up table. The Caughey and Thomas expression [21] is used here to parameterize the mobility dependence on the strength of the electric field parallel to the current flow.

The energy density losses W_{np} are defined by the interaction of carriers with the lattice. As mentioned before the leading phenomena in the description of the carriers scattering physics in the CMOS Silicon Photomultiplier are the impact ionization and the lattice and Auger recombination. It is possible to write the energy losses as:

$$W_n = \frac{3}{2}n\frac{k\,(T_n - T_L)}{\tau_n}\lambda_n + \frac{3}{2}kT_n\lambda_n R + E_g\left(G_n - R_n^A\right) \qquad (8.11)$$

$$W_p = \frac{3}{2}n\frac{k\,(T_p - T_L)}{\tau_p}\lambda_p + \frac{3}{2}kT_p\lambda_p R + E_g\left(G_p - R_p^A\right) \qquad (8.12)$$

where:

R^A is the Auger recombination rate
T_L is the lattice temperature
τ_n is the relaxation time for electrons
τ_p is the relaxation time for holes
E_g is the band-gap energy.

The relaxation times are also important parameters in the modelling of the CMOS SiPM detection structure. We use the approximation $\tau_n \approx \tau_p \approx 1$ ps, which is retrieved in accordance with the experimental conditions. In the heavy doped region the decreased bandgap separation is included by lowering the conduction band level by the same amount as the valence band is raised. The separation between valence and conduction band in this regime is modelled according to the experimental results and theoretical studies in [22].

8.3.2 Modeling of the Sensor Avalanche Breakdown Structure

The development of the semiconductor structure within CMOS technology required preferably to use also appropriate simulation framework. One of such simulation framework for CMOS technology is Silvaco TCAD [23], what we choose for simulations. Actually the basis of the theoretical description of the physical processes above is implemented in the Silvaco TCAD and allowed to optimize of many parameters according particular requirements. Hence our study we consider to perform on base of standard simulation framework Silvaco TCAD with detailed tests and

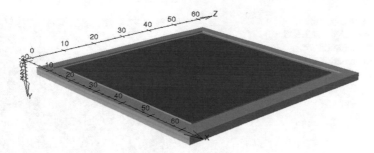

Fig. 8.4 The simulated 3D CMOS Silicon Photomultiplier sensor avalanche breakdown structure. In red the heavily doped p-region, in green the n-well. Around the p-region the STI are formed. The STI volume is invisible to simplify its identification in the 3D plot

optimisation of the parameters for a realistic and precise description of the of the sensor avalanche breakdown micro-cells.

We use the 3D option of Silvaco TCAD simulation framework for as mentioned above of importance of the shape of the pn-junction and electric field in particular areas. The size of the sensitive structure is $60 \times 60 \times 5 \,\mu m^3$. The general view of simulation domain is shown on Fig. 8.4. The structures under study are designed following the rules of CMOS technology $0.180 \,\mu m$. The sensor avalanche breakdown micro-cell is formed by planar p^{++} on n junction and following of the CMOS PDK rules.

We would like mentioned that the design of particular CMOS structures on a standard CMOS technology facility is not trivial because often the manufacturer does not provide an information about used material and the technology parameters. This happens in particular if the production and the R&D is based on Multi Project Wafer. It is hence a challenge for the modelling to identify the best profile fitting with the used technology and with the reduced set of technology information.

The cross-sectional view of one of the simplest possible CMOS avalanche break-down structure is shown on Fig. 8.5 and the doping profile at $x = 30 \,\mu m$ is shown on Fig. 8.6 for the simplicity the doping profile is represented by Gaussian distributions. The details of the doping profile are visible. The red colour corresponds to the heavily doped p^{++}-region. The green colour is the light-doped n-region.

The doping concentration of the n-epitaxial region is about $10^{15} \ cm^{-3}$. The n-well doping ranges from $10^{17} \ cm^{-3}$ in the space charge region to about $3 \times 10^{17} \ cm^{-3}$ at a depth of $0.41 \,\mu m$. The $4p^+$ doping peaks at $3 \times 10^{20} \ cm^{-3}$ at a depth of $0.1 \,\mu m$. The junction is at a depth of 142 nm.

The computational domain is described by a tetrahedral mesh defined with an adaptive method in correspondence with the initial doping profile. The number of nodes is 4×10^7, the maximal number allowed in Silvaco TCAD. The smaller mesh size is $0.1 \,\mu m$. This selection allows for a robust and convergent solution of the numerical problem. We verified that the numerical error in the estimation of the currents is in this case less than 3%.

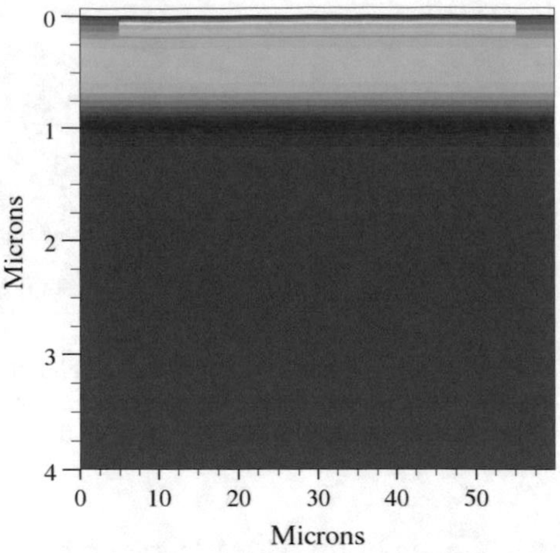

Fig. 8.5 Cross-sectional view of the simplest CMOS Silicon Photomultiplier avalanche breakdown structure

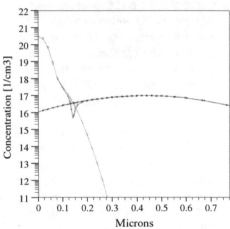

Fig. 8.6 Doping profile of the CMOS Silicon Photomultiplier structure at $x = 30\,\mu m$

8.3.3 Analysis of the Electric Field and Breakdown Conditions

A critical feature in the design of the CMOS SiPM is the possibility to obtain a uniform avalanche-breakdown mode along the sensitive area of the sensor. The condition is set by (8.4), which states that the ionization coefficients of the carriers and the corresponding current densities should be approximately constant along the junction defining the sensitive area of the device. As these quantities depend on the electric field, (8.1) relates them ultimately to the geometry, size and doping of the detection structure. In particular it is enough to calculate the electric field satisfying the breakdown condition and to verify that the electric field is constant along the active area of the sensor.

Fig. 8.7 Electric field profile of a SiPM detection structure without guard rings at 0.1 V above breakdown. At the bottom of the figure the zoomed area of the junction corner is visualized. The electric field strength at the borders of the junction is higher. It causes a local breakdown deteriorating the overall performance of the sensor

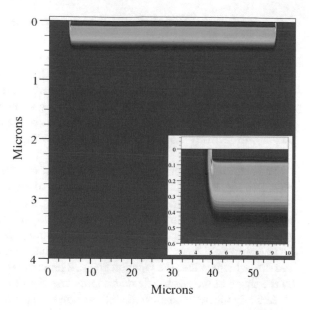

On Fig. 8.7 we show the electric field profile of a $p^{++}n$, correspondent to the structure on Fig. 8.5 at 0.1 V above breakdown. The analysis shows the strong non uniformity of the electric field in particular at the border and the corner of the *pn*-junction, the electric field rises abruptly of one order of magnitude. This behaviour is strongly disturb the breakdown conditions over the area of the *pn*-junction and build the local breakdown areas and deteriorates the performance (in particular efficiency) of the sensor, the high intrinsic gain of amplification will be only in this small areas on the periphery of the *pn*-junction. This analysis shows that development of the avalanche breakdown structures for the Silicon Photomultipliers is required special approach with detailed analysis.

8.4 Development of the Avalanche Breakdown Micro-cells

As stated before, the CMOS technology offers few possibilities of avoiding the local breakdown problem using only the available standard processes. The mathematical modelling is in this respect a powerful analysis and planning tool, as it gives an insight of the electric field of the structure, which is not experimentally observable. This information plays a central role in the correct design and is of guidance for the choice of the technological parameters.

8.5   Avalanche Breakdown Structure with STI Guard Ring

We first analyse the possibility offered by the using the shallow trench isolation. The standard CMOS Technology provided the STI around active area and allowed exclude the edges of the *pn*-junctions with disturbed areas of the electric fields. The structure with STI, structure with STI guard ring is shown on Fig. 8.8.

Around the junction STI is formed and filled with SiO_2. The width and depth of STI are respectively 0.5 μm and 0.4 μm, following the indications of the standard CMOS facility.

The electric field corresponding to the detection structure analyzed in Fig. 8.8 is shown on Fig. 8.9 in 3D view. The cross-sectional view of the electric field is shown on Fig. 8.10. The electric field is constant along the sensitive area of the avalanche breakdown structure. In comparison with Fig. 8.7, the electric field does not exhibit localized hot regions at the edges of the structure, due to the presence of the STI. The electric field has a peak value of approximately 4.6×10^5 V/cm at a depth of 142 nm, which defines the junction depth. The total width of the structure is defined by the shape of the electric field as approximately 420 μm.

Actually the mathematical modeling shows the principal possibility to design the sensor avalanche breakdown structure with STI guard ring and was chosen for the experimental investigations.

8.5.1   Avalanche Breakdown Structure with Virtual Guard Ring

Another possibility is offered by the virtual guard ring option. The cross-sectional view of the doping profile of such modelled structure is shown on Fig. 8.11. The

Fig. 8.8 Cross-sectional view of the 3D CMOS Silicon Photomultiplier structure. The total concentration as well as the STI are visible

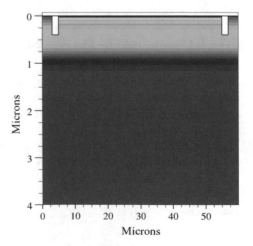

Fig. 8.9 Electric field of the SiPM detection structure shown on Fig. 8.4 at 0.1 V above breakdown. The electric field is constant along the sensitive area of the detector

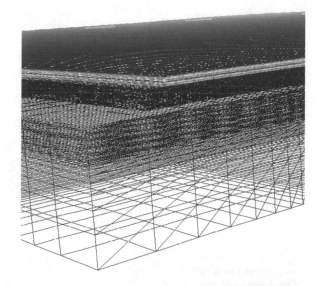

Fig. 8.10 2D electric field cross-sectional view of the SiPM detection structure shown on Fig. 8.4 at 0.1 V above breakdown. The electric field is constant along the sensitive area of the detector (Color figure online)

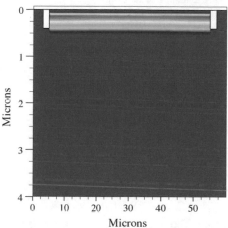

doping concentration of the n-epitaxial region is about 10^{15} cm^{-3}. The n-well doping ranges from 10^{17} cm^{-3} in the space charge region to about 3×10^{17} cm^{-3} at a depth of 0.41 μm. The p^+ doping peaks at 3×10^{20} cm^{-3} at a depth of 0.1 μm. The p^+ doping extends 1 μm longer on all sides of the sensitive area. The junction is at a depth of 142 nm. In addition, along the sides of the junction a p-well doped area with width 1 μm and with concentration ranging from 10^{17} cm^{-3} to about 3×10^{17} cm^{-3} at a depth of 0.41 μm is present. The STI isolation trench is positioned outside the active area and for simplicity not shown on the structure.

The electric field cross-sectional view is shown respectively on Fig. 8.12. The electric field is constant along the sensitive area of the detector. The electric field has a value of approximately 4.6×10^5 V/cm at a depth of 142 nm, defining the junction

Fig. 8.11 Cross-sectional view of the doping concentration of the SiPM detection structure with virtual guard rings. The electric field is constant along the sensitive area of the detector (red colour)

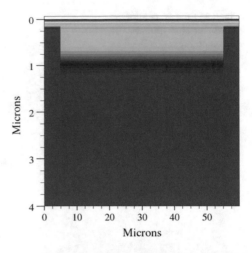

Fig. 8.12 Electric field cross-sectional view of the SiPM detection structure with virtual guard rings at 0.1 V above breakdown

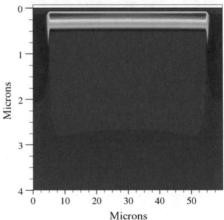

depth. The total width of the structure is defined by the shape of the electric field as approximately 420 μm. We observe that the virtual guard ring structure preserves the condition of the electric field at the edges of the junction, avoiding the occurrence of local breakdown and form very uniform of the electric field across the sensitive area of the avalanche breakdown micro-cell.

8.5.2 Analysis of the Current-Voltage Characteristics of Avalanche Breakdown Structures

The experimental observables of the mathematical model consist of quantities which can be measured with precision using dedicated measurement systems. The set of

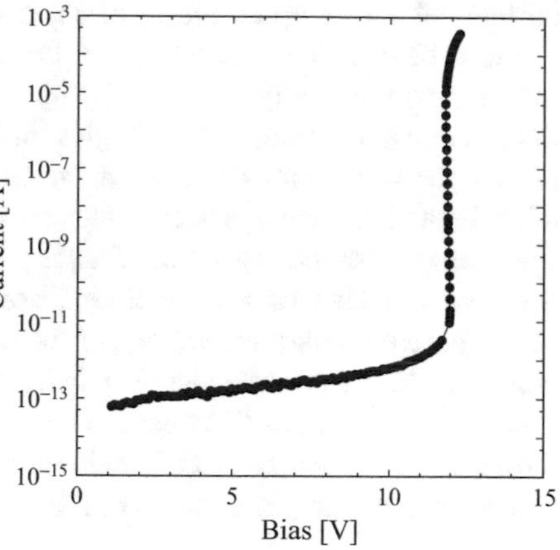

Fig. 8.13 Calculated current-voltage characteristics of a CMOS SIPM detection structure with STI guard rings

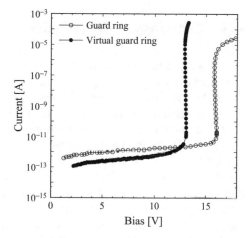

Fig. 8.14 Calculated current-voltage characteristics of a CMOS SIPM detection structure with virtual guard rings

measurements traditionally provided for the analysis of the semiconductor structures includes the dependence of the current of the detection structure corresponding with the applied bias.

The calculated current voltage characteristics of the CMOS SiPM detection structure with STI guard rings is shown on Fig. 8.13. The filled circles correspond to the calculation using the full model of the physical processes available in the Silvaco TCAD, including impact ionization. The curve follows the leakage current behaviour up to a level of approximately 1 pA. After that the electric field strength begins to be sizeable and the impact ionization process let the current increase exponentially and abruptly up to a level of approximately 10 μA, at which the external quenching resistor begins limiting the current of the device. The expected breakdown voltage of the sensor micro-cell is approximately 12 V.

The calculated current voltage characteristics of the CMOS SiPM detection structure with virtual guard rings is shown on Fig. 8.14. The curve follows the leakage current behaviour up to a level of approximately 1 pA. After that the electric field strength begins to be sizeable and the impact ionization process let the current increase exponentially and abruptly up to a level of approximately $10\,\mu$A, at which the external quenching resistor begins limiting the current of the device. On the same picture shown also the current-voltage correspond to the guard ring areas, the value of the breakdown voltage is 16 V. Clear seen that the breakdown voltage for the guard ring area is significantly higher. This is prevent of undesired avalanches processes in this region under working conditions, correspondent to the main area of micro-cell. The expected breakdown voltage of the structure with virtual guard ring is approximately 13 V. We observe that the expected level of leakage current is slightly higher than in the STI guard rings case. The higher breakdown voltage is consistent with the expectation that the electric field with virtual guard rings increases less abruptly and is more regular, in particular in the vicinity of the edges.

8.6 Experimental Investigations

To investigate the possibilities to develop of the Silicon Photomultiplier within the standard CMOS processes, we produce avalanche breakdown detection structures with the two different kinds of guard rings proposed above and modelled in the previous section. The production is performed in the 0.180 BCDLite IC MPW node at Global Foundries. We choose the process with n-type epitaxial layer, on which the photo-detection structures based on p^+/n junctions are formed. The active area of the junction is $50 \times 50\ \mu$m^2. We do not optimize the optical window above the active area with anti reflective coating. However, we minimize the thickness of the SiO$_2$ covering by using the minimal number of three metal layer offered in this technology. In fact, although there is clearly no metal above the active area of the sensor, in CMOS MPW the SiO$_2$ isolation layers between metal layers are deposited indistinctly on the whole area of the chip.

The structures are operated in breakdown mode and a quenching element is needed in order to stop the avalanche. We choose to implement a passive quenching element consisting of a resistor integrated around the sensitive cell. The used CMOS MPW line includes High Resistive Polysilicon as a standard process. We design a quenching element of 250 kΩ with a sheet resistivity of 2 kΩ and a minimal width of 0.5 μm. We produced a series of test resistors on the same wafer of the detection structures and we measured the spread of the resistor values to be 3% in this technology. This result is compatible with the requirements of the SiPM design.

8.6.1 Current-Voltage Characteristics

The current-voltage characteristics of the two produced structures is shown on Fig. 8.15. The structure with STI guard ring exhibits a slightly lower breakdown

Fig. 8.15 Measured current-voltage characteristics of the SiPM structures. The circles correspond to STI guard ring, the triangles to the virtual guard ring structures

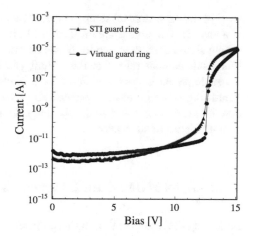

voltage (about 12.5 V) with respect to the structure with virtual guard ring (about 13.2 V). In comparison with the modelling results, the mathematical model and the experimental results are in good agreement regarding the prediction of the breakdown voltage, but the behaviour of the structure with STI guard ring is significantly different.

The dark current level is significantly different between the two structures. The virtual guard ring structure shows a subpicoampere dark current before the avalanche breakdown. At breakdown the current abruptly rises up to a few nanoamperes level. After breakdown rises linearly as it is limited by the quenching resistor. The structures with STI guard ring exhibits a larger dark current, reaching a few nanoamperes already before breakdown and a few microamperes at breakdown, before getting limited by the quenching resistor. The raising of the dark current obtained in STI guard rings is related as we suppose to defect of the material generation during the trench etch process. And in the conditions with high electric field the effect is dramatically increasing. As we already tested in other CMOS technology facilities, this effect is by far limiting the possibility of the use of STI guard rings in Multi Project Wafer [13]. The production of devices in MPW follows the standard processes and it is not possible a user-oriented optimization. As clear from a direct comparison of Fig. 8.15 and Figs. 8.13–8.14, the effect of the defect generation in the trenches process is not included in the mathematical model and further improvements are needed in order to predict the deterioration of the current properties using the STI technology in the production of the sensor.

Thermal e/h pairs generated in the SiPM sensitive area give rise to an output signal which is not distinguishable from a single detected photon. Such effect is called dark rate. We measure the amplitude of dark rate pulses in the two proposed photon-detection structures by biasing the structure at a voltage 0.2 V above the breakdown and observing the output signal with a 4 GHz bandwidth oscilloscope after an high frequency 18× voltage amplification stage. The virtual guard ring structure exhibits a dark rate signal with amplitude of approximately 20 mV, well above the 2 mV noise

level of the experimental setup. The dark rate measured for signals whose amplitude
is above the 0.5 single photon threshold (10 mV) is approximately 3 KHz. As the
output signal corresponding to the single photon is equivalent to the dark rate pulse,
this result demonstrates that the virtual guard ring structure is capable to provide
a sizable signal in correspondence of a photon detected. On the contrary, the STI
guard ring structure did not provide any dark rate signal in the same experimental
conditions. This result reflects the large leakage observed in the current voltage
characteristics of such structure.

8.6.2 Optical and Gain Characteristics

Following the results of the single cell performances, we produce on the same MPW
a prototype of SiPM consisting of an array of photo-detection cells with virtual
guard rings. For the photodetector characterization a 850 nm LED light source and a
circular lensed fiber with 10 μm spot size was used for injecting light into the SiPM.

The dc characterization is performed on wafer. The LED light source is operated
in continuous mode. The current-voltage characteristics of the SiPM prototype at
dark conditions and with light are shown on Fig. 8.16. The dependence of the current
versus voltage is following the behavior of the single cell and there is no significant
deviations from the expectations. The level of the current-voltage curve under light
exposure shows a visible increase, due to the current generated by photon detection.
Although the optical window of this prototype is not optimized the prototype exhibit
a sizable photon detection efficiency.

The photodetector characterization is performed with packaged samples of the
prototypes. The LED light source is pulsed with a pulse width of 10 ns. The SiPM
output signal is amplified with a high frequency 18× voltage amplifier and integrated
with the CAEN V1180 QDC within an integration gate of 150 ns.

Fig. 8.16 Measured
current-voltage
characteristics of an array of
5 × 5 SiPM structures with
virtual guard rings. The
crosses correspond to dark
condition, the triangles to
illumination condition

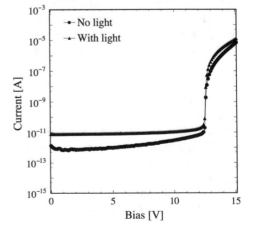

Fig. 8.17 Low photon flux spectrum detected by a SiPM prototype

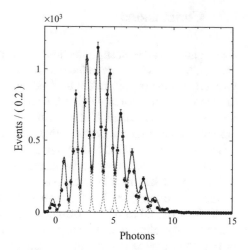

Figure 8.17 shows the spectrum of the signal corresponding to the used low photon flux source. The histogram plots the distribution of the integrated sensor current for an average detected flux of photons $\mu = 4$. Here the sensor is biased at 1 V above the breakdown voltage. The integrated current is shown in units of correspondent number of photons. The histogram exhibits a series of peaks at etc. which we identify as induced by 0, 1, 2, 3 etc. photons.

The 0-photon peak corresponds to a period on which there is no photo-induced avalanche, only electric noise of the external circuit components. The black continuous thick line on Fig. 8.17 shows a fit to the data assuming a poissonian distribution of the incident LED light pulses. We fitted the spectrum with a sum of Gaussians, with area weighted according to a poissonian distribution, using a fitting technique explained in [13]. The peaks are assumed to be equally spaced. The spectrum is consistent with a poisson-distributed light and the histograms on Fig. 8.3 and Fig. 8.4 can be interpreted as a representation of a specific photon number state.

The fitted distance between the peaks is $G = (1.706 \pm 0.003) \times 10^6$ electrons. This distance represents the number of electrons generated during the avalanche process before the quenching occurs. In other words it is the intrinsic gain of the sensor. We observe that the intrinsic gain is high enough to separate the signal corresponding to the single photon detection from the electronic noise level. The possibility of observing the single photon spectrum on Fig. 8.17 depends strongly on the uniformity of the gain of each sensor microcell and is a proof of the high quality and reliability of the CMOS technology processes used in the fabrication.

8.7 Conclusions

The progress on the mass production even for the conventional SiPMs is fully laying on the CMOS technology industrial facilities to optimise the cost and performance. Furthermore the overall future development of the new SiPM structures, as Digital Silicon Photomultipliers, Digital Silicon Photomultiplier imagers, Avalanche Pixel Structures for the detection of ionization particles is dependent on the CMOS technology. We have shown evidence that the main component—the photosensor—can be realized in the CMOS technology. Further developments will concentrate on the optimisation of the sensor performance, using the modern development in CMOS technology as 3D Inter Connection technology, which is very efficient and perspective for the sensor part. The future main effort will be focused on the development of the fully digital readout electronics and the additional powerful processing electronics on the chip. It will widely improve the overall performance and reliability of the SiPMs, as sensor for the low photon flux and widely increase the areas of applications [25].

References

1. H. Toshikaza, *Photomultiplier Tubes, Basics and Applications* (Hamamatsu Photonics K. K., Electron Tube Division, Japan, 2006)
2. M. Atac, J. Park, D. Cline, D. Chrisman, M. Petroff, E. Anderson, Nucl. Instrum. Methods Phys. Res. A **314**, 56 (1992)
3. V. Saveliev, V. Golovin, Nucl. Instrum. Methods A **442**, 223 (2000)
4. V. Golovin, V. Saveliev, Nucl. Instrum. Methods A **518**, 560 (2004)
5. A.G. Chynoweth, K.G. McKay, Phys. Rev. **102**, 369 (1956)
6. V. Saveliev, *Quantum Detector Arrays* (US Patent US 7,825,384, 2010)
7. N. D'Ascenzo, V. Saveliev, Q. Xie, L. Wang, in *Optoelectronics—Materials and Devices*, ed. by S. Pyshkin, J. Ballato (Intech, 2015)
8. N. D'Ascenzo, P. Marrocchesi, C.S. Moon, F. Morsani, L. Ratti, V. Saveliev, A. Savoy-Navarro, Q. Xie, J. Instrum. **9**, C03027
9. N. D'Ascenzo, V. Saveliev, in *Photodetectors*, ed. by S. Pyshkin, J. Ballato (Intech, 2015)
10. D.J. Herbert, V. Saveliev, N. Belcari, N. D'Ascenzo, A. Del Guerra, A. Golovin, I.E.E.E. Trans, Nucl. Sci. **53**, 389 (2006)
11. V. Saveliev, in *Advances Optical and Photonic Devices*, ed. by W. Shi (Intech, 2012)
12. N. D'Ascenzo, V. Saveliev, in *Photodetectors*, ed. by J.-W. Shi (Intech, 2012)
13. N. D'Ascenzo, V. Saveliev, L. Wang, Q. Xie, J. Instrum. **10**, C08017 (2015)
14. M. Lee, H. Rucker, W. Choi, I.E.E.E. Electr, Dev. Lett. **33**, 80 (2012)
15. M. Lee, H. Rucker, W. Choi, I.E.E.E. Electr. Dev. Lett. **37**, 60 (2016)
16. N. Izhaky, M.T. Morse, S. Kohel, O. Cohen, D. Rubin, A. Barkai, G. Sarid, R. Cohen, M.J. Paniccia, I.E.E.E.J. Sel, Top. Quantum Electron. **12**, 1688 (2006)
17. W. Sul, J. Oh, C. Lee, G. Cho, W. Lee, S. Kim, J. Rhee, I.E.E.E. Electr, Dev. Lett. **31**, 41 (2010)
18. K. Katayama, T. Toyabe, *IEDM Technical Digest* (1989), p. 135
19. W. Shockley, W.T. Read, Phys. Rev. **87**, 835 (1952)
20. J.G. Fossum, R.P. Mertens, D.S. Lee, J.F. Nijs, Solid State Electron. **26**, 569 (1983)
21. D.M. Caughey, R.E. Thomas, Proc. IEEE **55**, 2192 (1967)
22. J.W. Slotboom, H.C. De Graaf, Solid State Electron. **19**, 857 (1976)

23. Silvaco. www.silvaco.com
24. N. D'Ascenzo, V. Saveliev, Q. Xie, *Proceedings of the 4th International Conference on Photonics, Optics and Laser Technology (PHOTOPTICS 2016)*, p. 215
25. N. D'Ascenzo, V. Saveliev, Nucl. Instrum. Methods A **695**, 265 (2012)

Chapter 9
Coherent Hemodynamics Spectroscopy: A New Technique to Characterize the Dynamics of Blood Perfusion and Oxygenation in Tissue

Sergio Fantini, Kristen T. Tgavalekos, Xuan Zang and Angelo Sassaroli

Abstract Hemodynamic-based neuroimaging techniques such as near-infrared spectroscopy (NIRS) and functional magnetic resonance imaging (fMRI) are directly sensitive to the blood volume fraction and oxygen saturation of blood in the probed tissue. The ability to translate such hemodynamic and oxygenation measurements into physiological quantities is critically important to enhance the effectiveness of NIRS and fMRI in a broad range of applications aimed at medical diagnostic or functional assessment. Coherent hemodynamics spectroscopy (CHS) is a novel technique based on the measurement (with techniques such as NIRS or fMRI) and quantitative analysis (with a novel mathematical model) of coherent hemodynamics in living tissues. Methods to induce coherent hemodynamics in humans include controlled perturbations to the mean arterial pressure by paced breathing or by timed inflations of pneumatic cuffs wrapped around the subject's legs. A mathematical model recently outlined translates coherent hemodynamics into physiological measures of the capillary and venous blood transit times, cerebral autoregulation, and cerebral blood flow. A typical method to analyze the optical signal from non-invasive NIRS measurements of the human brain is the modified Beer-Lambert law (mBLL), which does not allow the discrimination of hemodynamics taking place in the scalp and skull from those occurring in the brain cortex. A hybrid method using continuous wave NIRS (with the mBLL) together with frequency-domain NIRS (with a two-layer diffusion model) was successfully used to discriminate oscillatory hemodynamics in the superficial (extracerebral) tissue layer from that in deeper, cerebral tissue.

S. Fantini (✉) · K. T. Tgavalekos · X. Zang · A. Sassaroli
Department of Biomedical Engineering, Tufts University, 4 Colby Street, Medford, MA 02155, USA
e-mail: sergio.fantini@tufts.edu

P. A. Ribeiro and M. Raposo (eds.), *Optics, Photonics and Laser Technology*, Springer Series in Optical Sciences 218, https://doi.org/10.1007/978-3-319-98548-0_9

9.1 Hemodynamic-Based Neuroimaging Techniques

With a few exceptions (for example evoked potential tests), functional neuroimaging techniques do not directly measure neural activation; they are sensitive to hemodynamic and metabolic changes associated with brain activity. For example near infrared spectroscopy (NIRS) and functional magnetic resonance imaging (fMRI) are two of the prominent hemodynamic-based neuroimaging techniques. The first one measures the concentrations of oxy- and deoxy-hemoglobin in brain tissue, whereas the latter measures a blood oxygenation level dependent (BOLD) signal that is mostly associated with the concentration of deoxy-hemoglobin in brain tissue and with cerebral blood volume. To assess brain function with these neuroimaging techniques, one needs to characterize the relationship between neural activation and hemodynamic responses, as well as how these hemodynamic responses translate into the measured signals. Noteworthy hemodynamic models proposed in the literature to address these questions include the oxygen diffusion limitation model [1] and a windkessel model [2]. The former model was proposed to account for the large imbalance, observed with positron emission tomography (PET), between blood flow and oxygen consumption changes associated with brain activation. The latter model was introduced to specify the relationship between the dynamics of blood flow and blood volume. These general models were pioneering contributions whose approaches can be adapted to any hemodynamic-based neuroimaging techniques.

Near-infrared spectroscopy (NIRS) is a non-invasive optical method that operates in the 600-900 nm wavelength diagnostic spectral window range. In this wavelength range, oxy-hemoglobin and deoxy-hemoglobin are the main absorbers in tissue, together with some absorption contributions from water and lipids. NIRS has found a variety of applications to the brain, both for functional studies [3–5] and for clinical feasibility [6] in areas such as anesthesiology [7], neurological critical care [8, 9], and electroconvulsive therapy [10]. The spatial and temporal correlations between the BOLD fMRI and the optical NIRS signals elicited by brain activation have been demonstrated in a number of studies [11–14]. Several hemodynamic models have been proposed in the NIRS field, most of them requiring the solution of complex systems of differential equations with many unknown parameters [15–18]. We have recently developed a mathematical model to relate the tissue concentration and saturation of hemoglobin, as measured with NIRS and fMRI, to blood flow, blood volume, and oxidative metabolism [19, 20]. This mathematical model is analytical, and therefore relatively simple and computationally inexpensive.

Recently, we have introduced a novel technique, coherent hemodynamics spectroscopy (CHS), for quantitative studies of tissue hemodynamics based on data collected with NIRS or fMRI [19]. The CHS technique is based on the application of our mathematical model to measured coherent hemodynamics, which may be occurring spontaneously or may be induced by controlled systemic perturbations to the mean arterial blood pressure. In Sect. 9.2, we describe the basic ideas and the general formulation of our mathematical hemodynamic model and of the new CHS technique.

In NIRS, the relative dynamics of oxy- and deoxy-hemoglobin concentrations are usually obtained from the intensity signals measured at a minimum of two wavelengths by applying the modified Beer-Lambert law (mBLL) [21, 22]. Unlike diffusion theory, the mBLL, which has also been used to investigate photon migration in tissues, does not require highly scattering conditions. Nevertheless, the mBLL is fundamented on two key assumptions: (1) the changes in optical intensity are only due to tissue absorption changes; (2) the absorption changes are uniformly distributed within tissue. While the first assumption is often fulfilled to a good approximation, the second one is more questionable and may be violated in a number of cases. For example, brain activation elicited in a number of functional protocols is spatially localized, and so are the associated hemodynamic changes measured with NIRS. Hence, it would be relevant to release the homogeneous absorption changes assumption and instead use a light propagation model that considers (at least partially) the human head complex geometry and the heterogeneous hemodynamics associated with in vivo brain perfusion.

In addition to the physiological heterogeneity intrinsic of living tissue, there is also an issue of anatomical heterogeneity, which is particularly critical in the case of non-invasive NIRS measurements that aim to sense brain tissue with an optical probe placed on the scalp surface. To a first approximation, the human head may be represented by a set of tissue layers such as scalp, skull, subarachnoid space, brain cortex, etc. In the literature, a two-layered diffusion model has been used for measuring the baseline optical properties of two effective tissue layers, the first one representing the scalp and skull (the extracerebral tissue), and the second one representing the brain [23–25]. Short source detector separations (~0.5 cm) can be introduced in NIRS to provide a reference signal that is representative of the superficial hemodynamics in the scalp and skull, and that is used to regress out superficial-tissue contributions from the optical signal measured at longer (~3 cm) source-detector separations [26–29]. Here, we apply a two-layer diffusion model to multi-distance, frequency-domain NIRS data to estimate the baseline optical properties of extra-cerebral and cerebral tissue layers, as well as the thickness of the extra-cerebral layer. Then, we use phasor analysis, in conjunction with an estimate of the mean partial optical pathlength in the two layers, to assess the induced oscillatory hemodynamics (frequency: 0.059 Hz in this work) in the extra-cerebral and cerebral tissues from continuous wave NIRS measurements at short (8 mm) and long (38 mm) source-detector separations. The two-layer diffusion model and our approach to two-layer hemodynamics assessment are outlined in Sect. 9.3, experimental methods are described in Sect. 9.4, and representative results on a human subject are reported in Sect. 9.5.

9.2 Coherent Hemodynamics Spectroscopy for the Quantitative Assessment of Cerebral Hemodynamics

9.2.1 Notation for the Tissue Concentrations of Oxy- and Deoxy-Hemoglobin

As we have seen above, NIRS and fMRI are mostly sensitive to the concentration and oxygen saturation of hemoglobin in blood-perfused tissues. Therefore, to quantitatively describe the tissue hemodynamics as measured by NIRS and fMRI, one needs to consider the temporal evolution of the concentrations of oxy-hemoglobin and deoxy-hemoglobin in tissue. The standard notation for oxy- and deoxy-hemoglobin concentrations is, respectively, [HbO$_2$] and [Hb], or [O$_2$Hb] and [HHb], but we will use here a single-letter notation of O and D. Dynamic changes in O and D are considered in the time domain and in the frequency domain. In the time domain, the temporal evolution of changes from baseline or steady-state values (O_0 and D_0) are indicated as $\Delta O(t)$ and $\Delta D(t)$. In the frequency domain, sinusoidal changes at any frequency ω are indicated with bold-face phasor notation, $\mathbf{O}(\omega)$ and $\mathbf{D}(\omega)$, where these phasors are two-dimensional vectors whose amplitude and phase represent the amplitude and phase of the corresponding sinusoidal oscillations [30]. Representative time traces of $\Delta O(t)$ and $\Delta D(t)$ are reported in a general case of arbitrary temporal dynamics in Fig. 9.1a, and in a case of quasi sinusoidal oscillations in Fig. 9.1b. The inset of Fig. 9.1b shows the phasor representation of the oscillatory hemodynamics of oxy- and deoxy-hemoglobin concentrations.

Because hemoglobin concentration dynamics are driven by perturbations in three physiological quantities, namely blood volume, blood flow, and metabolic rate of oxygen, we introduce subscripts V, F, and $\dot{O}$ to indicate the contributions to O and D dynamics from these three physiological sources, respectively. A lack of subscript indicates the overall hemoglobin concentration dynamics resulting from all physiological sources.

We introduce a symbol S_V for the average oxygen saturation of the hemoglobin in the vascular compartment(s) whose volume fraction features dynamic changes. S_V is defined as follows:

$$S_V = \frac{\Delta O_V(t)}{\Delta O_V(t) + \Delta D_V(t)}, \tag{9.1}$$

where S_V is considered to be a constant under the assumption that the volume-varying vascular compartment has a well-defined and constant average saturation, at least over time scales of typical observation times.

It has been previously shown that techniques that are solely sensitive to the tissue concentrations of oxy- and deoxy-hemoglobin (such as NIRS and fMRI) are unable to discriminate the contributions of blood flow and metabolic rate of oxygen to the dynamic changes in O and D [20]. Therefore, we consolidate dynamic changes

Fig. 9.1 Representative time traces of dynamic changes in oxy-hemoglobin and deoxy-hemoglobin concentrations ($\Delta O(t)$ and $\Delta D(t)$). **a** Arbitrary dynamics, **b** oscillatory dynamics at a well-defined frequency. The inset of panel (**b**) shows the phasor representation of the quasi sinusoidal oscillations of oxy- and deoxy-hemoglobin concentrations (**O** and **D**)

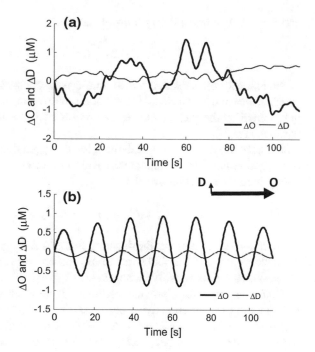

due to blood flow and metabolic rate of oxygen according to the following notation: $\Delta O_{F,\dot{O}}(t)$ and $\Delta D_{F,O}(t)$ in the time domain, $\mathbf{O}_{F,\dot{O}}(\omega)$ and $\mathbf{D}_{F,\dot{O}}(\omega)$ in the frequency domain.

We also introduce notation for the physiological dynamics associated with cerebral blood volume (CBV), cerebral blood flow (CBF), and cerebral metabolic rate of oxygen (CMRO$_2$). We use lower-case notation to indicate the dimensionless, relative dynamics with respect to baseline values (in the time domain) or average values (in the frequency domain), which are indicated with a 0 subscript. Explicitly, $cbv(t) = (CBV(t) - CBV_0)/CBV_0$, $cbf(t) = (CBF(t) - CBF_0)/CBF_0$, $cmro_2(t) = (CMRO_2(t) - CMRO_{2|0})/CMRO_{2|0}$ in the time domain, and $\mathbf{cbv}(\omega) = \mathbf{CBV}(\omega)/CBV_0$, $\mathbf{cbf}(\omega) = \mathbf{CBF}(\omega)/CBF_0$, $\mathbf{cmro}_2(\omega) = \mathbf{CMRO}_2(\omega)/CMRO_{2|0}$, in the frequency domain.

9.2.2 Time Domain Representation

According to the notation introduced in Sect. 9.2.1, we can write:

$$\Delta O(t) = \Delta O_V(t) + \Delta O_{F,\dot{O}}(t), \tag{9.2}$$

$$\Delta D(t) = \Delta D_V(t) + \Delta D_{F,\dot{O}}(t), \tag{9.3}$$

and from (9.1) it immediately follows that:

$$(1 - S_V)\Delta O_V(t) = S_V \Delta D_V(t). \tag{9.4}$$

Equation (9.4) states that the volume-driven changes $\Delta O_V(t)$ and $\Delta D_V(t)$ are synchronous and proportional to each other, with a constant of proportionality that is determined by the average oxygen saturation of the volume-varying vascular compartment(s).

Because blood flow and metabolic rate of oxygen dynamics solely affect the rate of oxygen removal from hemoglobin, which in turn transform oxy-hemoglobin into deoxy-hemoglobin, it follows that:

$$\Delta O_{F,\dot{O}}(t) = -\Delta D_{F,\dot{O}}(t). \tag{9.5}$$

Equation (9.5) states that the dynamic changes $\Delta O_{F,\dot{O}}(t)$ and $\Delta D_{F,\dot{O}}(t)$ resulting from blood flow and metabolic rate of oxygen changes are also proportional to each other, with a proportionality factor of -1.

Equations (9.2)–(9.5) provide relationships between the overall oxy- and deoxy-hemoglobin concentration dynamics ($\Delta O(t)$, $\Delta D(t)$) and their components due to blood volume changes ($\Delta O_V(t)$, $\Delta D_V(t)$) or to blood flow and metabolic rate of oxygen changes $\left(\Delta O_{F,\dot{O}}(t), \Delta D_{F,\dot{O}}(t)\right)$. In order to translate dynamic measurements of hemoglobin concentration into measurements of the underlying changes in blood volume cbv(t)), blood flow (cbf(t)), and metabolic rate of oxygen (cmro$_2$(t)), one needs to specify the relationship between hemoglobin concentration and physiological quantities. This critical point is addressed in Sect. 9.2.4.1.

9.2.3 Frequency Domain Representation

The time-domain representation provided by (9.2)–(9.5) is expressed by the following phasor equations in the frequency domain:

$$\mathbf{O}(\omega) = \mathbf{O}_V(\omega) + \mathbf{O}_{F,\dot{O}}(\omega), \tag{9.6}$$

$$\mathbf{D}(\omega) = \mathbf{D}_V(\omega) + \mathbf{D}_{F,\dot{O}}(\omega), \tag{9.7}$$

$$(1 - S_V)\mathbf{O}_V(\omega) = S_V \mathbf{D}_V(\omega). \tag{9.8}$$

$$\mathbf{O}_{F,\dot{O}}(\omega) = -\mathbf{D}_{F,\dot{O}}(\omega). \tag{9.9}$$

Equation (9.8) states that the volume-driven oscillations $\mathbf{O}_V(\omega)$ and $\mathbf{D}_V(\omega)$ are in phase and their relative amplitude can be determined by the average oxygen saturation of the volume-oscillating vascular compartment(s). Equation (9.9) states that the dynamic oscillations $\mathbf{O}_{F,\dot{O}}(\omega)$ and $\mathbf{D}_{F,\dot{O}}(\omega)$ resulting from blood flow and metabolic rate of oxygen oscillations are in opposition of phase and have the same amplitude.

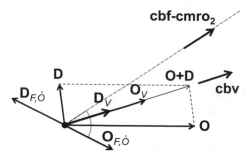

Fig. 9.2 Phasor decomposition of measured oscillations of oxy-hemoglobin (**O**) and deoxy-hemoglobin (**D**) concentrations, into their components due to blood volume (**O**$_V$ and **D**$_V$) and the combination of blood flow and oxygen consumption (**O**$_{F,\dot{O}}$ and **D**$_{F,\dot{O}}$) oscillations. The cerebral blood volume phasor (**cbv**) is in phase with **O**$_V$, whereas the blood flow and metabolic rate of oxygen phasor (**cbf** − **cmro**$_2$) leads **O**$_{F,\dot{O}}$ by the angle indicated in the figure (as described in Sect. 9.2.4.2)

We highlight the elegance of the phasor representation of (9.6)–(9.9) by illustrating how these equations can be solved graphically. The system of (9.6)–(9.9) is a set of four phasor equations, i.e. four 2-D vector equations, which correspond to eight scalar equations. Considering that **O**(ω) and **D**(ω) are measurable quantities (certainly both of them with NIRS), this set of equations contains nine unknowns: the two phasor components of each of **O**$_V$, **O**$_{F,\dot{O}}$, **D**$_V$, and **D**$_{F,\dot{O}}$, and the scalar quantity S_V. Therefore, the number of unknowns exceeds the number of equations by one [31]. To solve this system of equations one may assume a value for S_V, which has an upper limit given by the arterial saturation (say, 1) and a lower limit given by the lowest venous saturation (say, 0.5 under normal conditions at rest, but possibly lower under special conditions). In Fig. 9.2, we illustrate how to solve the system of (9.6)–(9.9) under the assumption that $S_V = 0.65$. The measured phasors **O** and **D** are combined by phasor addition into **O** + **D**, which represents the phasor of total hemoglobin concentration. Because of (9.9), only **O**$_V$ and **D**$_V$ contribute to the total hemoglobin concentration phasor (since **O**$_{F,\dot{O}}$ + **D**$_{F,\dot{O}}$ = 0), and (9.8) specifies that **O**$_V$ and **D**$_V$ are in phase with each other and therefore with **O** + **D**. As a result, **O**$_V$ and **D**$_V$ are aligned with **O** + **D** and the magnitude of **O**$_V$ is S_V (or 0.65) times the magnitude of **O** + **D** whereas **D**$_V$ is $1 - S_V$ (or 0.35) times **O** + **D**. From (9.6) and (9.7) it follows that **O**$_{F,\dot{O}}$ is the vector from the tip of **O**$_V$ to the tip of **O**, and **D**$_{F,\dot{O}}$ is the vector from the tip of **D**$_V$ to the tip of **D**. **O**$_{F,\dot{O}}$ and **D**$_{F,\dot{O}}$ are translated to start at the origin in Fig. 9.2, to show that they are opposite to each other, as per (9.9).

Figure 9.2 shows how, from the measured phasors (i.e. oscillations) of oxy- and deoxy-hemoglobin concentrations, one can find the phasor component associated with blood volume oscillations, and the phasor component associated with blood flow and metabolic rate of oxygen oscillations. Now, similar to the time-domain case, the key missing piece to translate oscillations of hemoglobin concentration into oscillations of physiological quantities is knowledge of the relationships between

$O_V(\omega)$ and $\mathbf{cbv}(\omega)$, and between $O_{F,\dot{O}}(\omega)$ and $\mathbf{cbf}(\omega)$, $\mathbf{cmro_2}(\omega)$. This critical point is addressed in Sect. 9.2.4.2.

9.2.4 A Hemodynamic Model to Relate Hemoglobin Concentrations to Physiological Quantities

A mathematical model has been recently developed to derive analytical relationships between the concentrations of oxy- and deoxy-hemoglobin concentrations in tissue, and the key physiological quantities (blood volume, blood flow, and metabolic rate of oxygen) that determine the steady-state values and drive dynamic changes in the concentrations of oxy- and deoxy-hemoglobin in living tissue [19, 20]. This model relies on the fact that every red blood cell, and the hemoglobin molecules it contains, must travel from an arterial compartment (which delivers blood to tissue), through a capillary compartment (including small arterioles, where oxygen exchange from blood to tissue occurs), to a venous compartment (which drains blood from tissue). Because oxygen diffusion from blood to tissue only occurs at the capillary level (here we include in the capillary compartment the smaller arterioles that are also involved with oxygen diffusion), the model treats the oxygen desaturation of blood according to just two parameters: the mean capillary transit time ($t^{(c)}$) and the rate constant of oxygen diffusion from capillary blood to tissue (α). Dynamic perturbations to either parameter (changes in blood flow velocity for $t^{(c)}$, changes in metabolic rate of oxygen for α) cause dynamic perturbations to the oxygen saturation of capillary blood, which propagate to the venous compartment according to another key model parameter, the venous blood transit time ($t^{(v)}$). We observe that while $t^{(c)}$ is a well-defined parameter, independent of the tissue volume sampled (as long as it includes the entire capillary compartment, which has linear dimensions of 1 mm or less), $t^{(v)}$ monotonically increases with the volume of probed tissue, because longer and longer venous drainage vessels are included in larger and larger probed volumes.

The model derives the dynamic effects of ideal step changes in blood volume, blood flow velocity, and metabolic rate of oxygen to the tissue concentrations of oxy- and deoxy-hemoglobin, and translates them to general perturbations with transfer function analysis, i.e. by considering the microvasculature as a linear, time-invariant system. In this approach, the microvascular system (for which dynamics in blood volume, blood flow, and metabolic rate of oxygen are the input, and the concentrations of oxy- and deoxy-hemoglobin are the output) is fully described by impulse response functions in the time domain, and by transfer functions in the frequency domain.

Because of the relationships between dynamic changes in oxy- and deoxy-hemoglobin, expressed by (9.2)–(9.5) in the time domain and by (9.6)–(9.9) in the frequency domain, we only need to specify the effects of blood volume, blood flow, and metabolic rate of oxygen on the concentration of one hemoglobin species, say oxy-hemoglobin. This is what we report in Sects. 9.2.4.1 for arbitrary hemodynamics, and in Sect. 9.2.4.2 for oscillatory hemodynamics.

9.2.4.1 Time-Domain Relationships

The contribution of oxy-hemoglobin concentration changes due to blood volume dynamics (ΔO_V) is written as follows [19]:

$$\Delta O_V(t) = \text{ctHb}\left[S^{(a)}\text{CBV}_0^{(a)}\text{cbv}^{(a)}(t) + \langle S^{(c)}\rangle \mathcal{F}^{(c)}\text{CBV}_0^{(c)}\text{cbv}^{(c)}(t) \right.$$
$$\left. + S^{(v)}\text{CBV}_0^{(v)}\text{cbv}^{(v)}(t) \right] \tag{9.10}$$

being ctHb the blood hemoglobin concentration, S is the oxygen saturation of hemoglobin, CBV_0 is the baseline blood volume, $\text{cbv}(t)$ is the relative blood volume change (with respect to baseline), $\mathcal{F}^{(c)}$ is the Fåhraeus factor (the ratio of capillary to large vessel hematocrit), and superscripts (a), (c), and (v) specify the arterial, capillary, and venous vascular compartments, respectively. Equation (9.10) shows that blood volume dynamics result in instantaneous changes in oxy-hemoglobin concentration, through proportionality factors expressed in terms of the concentration and oxygen saturation of hemoglobin in blood.

The contribution of oxy-hemoglobin concentration changes $(\Delta O_{F,\dot{o}})$ due to blood flow and metabolic rate of oxygen dynamics can be expressed as [19]:

$$\Delta O_{F,\dot{o}}(t)$$
$$= \text{ctHb}\left[\frac{\langle S^{(c)}\rangle}{S^{(v)}}\left(\langle S^{(c)}\rangle - S^{(v)}\right)\mathcal{F}^{(c)}\text{CBV}_0^{(c)}h_{RC-LP}^{(c)}(t) + \left(S^{(a)} - S^{(v)}\right)\text{CBV}_0^{(v)}h_{G-LP}^{(v)}(t) \right]$$
$$* (\text{cbf}(t) - \text{cmro}_2(t)) \tag{9.11}$$

The $*$ is here indicating a convolution product, $\text{cbf}(t)$ and $\text{cmro}_2(t)$ are the relative blood flow and metabolic oxygen rate changes (with respect to baseline), and $h_{RC-LP}^{(c)}(t)$ and $h_{G-LP}^{(v)}(t)$ are the impulse response functions that describe the low-pass (LP) effects related to the microcirculation in the capillary and venous vascular compartments, respectively.

The capillary impulse response function is approximated by that of a low-pass resistor-capacitor (RC) circuit equivalent:

$$h_{RC-LP}^{(c)}(t) = u(t)\frac{1}{\tau^{(c)}}e^{-t/\tau^{(c)}}, \tag{9.12}$$

where $u(t)$ is the unit step function (which is 0 for a negative argument, and 1 for a positive or zero argument) and $\tau^{(c)}$ is the capillary time constant $(\tau^{(c)})$ that is related to $t^{(c)}$ as follows:

$$\tau^{(c)} = \frac{t^{(c)}}{e}. \tag{9.13}$$

The venous impulse response function is approximated by that of a time-shifted Gaussian low-pass filter:

$$h_{G-LP}^{(v)}(t) = \frac{1}{t_r^{(v)}} e^{-\pi \left(t - t_{0.5}^{(v)}\right)^2 / \left(t_r^{(v)}\right)^2}, \tag{9.14}$$

where $t_{0.5}^{(v)}$ is the time-shift constant and $t_r^{(v)}$ is the time-width constant. These two time constant are close to each other, since $t_{0.5}^{(v)} = (5/6) t_r^{(v)}$. If we identify $t_{0.5}^{(v)}$ with the venous time constant ($\tau^{(v)}$), one finds that $\tau^{(v)}$ is the mean of the capillary and venous blood transit times:

$$\tau^{(v)} = \frac{t^{(c)} + t^{(v)}}{2}. \tag{9.15}$$

This treatment shows that any dynamic changes in blood flow or metabolic rate of oxygen result in delayed and temporally spread changes in the tissue concentrations of oxy- and deoxy-hemoglobin. These temporal effects occur on a time scale that is determined by the capillary and venous blood transit times (as specified by the capillary and venous time constants in (9.13) and (9.15)), and is analytically reflected in the convolution product of (9.11).

9.2.4.2 Frequency-Domain Relationships

The oxy-hemoglobin phasor ($\mathbf{O}_V$) that describes oxy-hemoglobin dynamics due to sinusoidal oscillations of blood volume at angular frequency ω is written as follows [19]:

$$\mathbf{O}_V(\omega) = \text{ctHb}\left[S^{(a)} \text{CBV}_0^{(a)} \mathbf{cbv}^{(a)}(\omega) + \langle S^{(c)} \rangle \mathcal{F}^{(c)} \mathbf{cbv}^{(c)}(\omega) \right.$$
$$\left. + S^{(v)} \text{CBV}_0^{(v)} \mathbf{cbv}^{(v)}(\omega) \right] \tag{9.16}$$

where $\mathbf{cbv}(\omega)$ is the relative blood volume phasor (normalized to the mean value of blood volume), and we use the same notation as in (9.10). Equation (9.16) shows that blood volume oscillations cause in-phase oscillations of oxy-hemoglobin concentration. In fact, $\mathbf{O}_V$ and the blood volume phasors are linked by a linear relationship with all real and positive coefficients that are expressed in terms of the concentration and oxygen saturation of hemoglobin in blood.

The oxy-hemoglobin phasor ($\mathbf{O}_{F,\dot{o}}$) that describes oxy-hemoglobin dynamics due to sinusoidal oscillations of blood flow and metabolic rate of oxygen is written as follows [19]:

$$\mathbf{O}_{F,\dot{o}}(\omega) = \text{ctHb}\left[\frac{\langle S^{(c)} \rangle}{S^{(v)}} \left(\langle S^{(c)} \rangle - S^{(v)} \right) \mathcal{F}^{(c)} \text{CBV}_0^{(c)} \mathcal{H}_{RC-LP}^{(c)}(\omega) \right.$$
$$\left. + \left(S^{(a)} - S^{(v)} \right) \text{CBV}_0^{(v)} \mathcal{H}_{G-LP}^{(v)}(\omega) \right] [\mathbf{cbf}(\omega) - \mathbf{cmro}_2(\omega)] \tag{9.17}$$

where $\mathbf{cbf}(\omega)$ and $\mathbf{cmro}_2(\omega)$ are the relative blood flow and metabolic rate of oxygen phasors (normalized to the corresponding mean values), and $\mathcal{H}^{(c)}_{RC-LP}(\omega)$ and $\mathcal{H}^{(v)}_{G-LP}(\omega)$ are the transfer functions that describe the low-pass (LP) effects associated with the microcirculation in the capillary and venous vascular compartments, respectively. Notice that the convolution product in the time domain (9.11) is replaced by a regular product in the frequency domain (9.17).

The capillary and venous transfer functions are given by the Fourier transform of the capillary and venous pulse response functions given in (9.12) and (9.14), respectively:

$$\mathcal{H}^{(c)}_{RC-LP}(\omega) = \frac{1}{1 + i\omega\tau^{(c)}}, \tag{9.18}$$

$$\mathcal{H}^{(v)}_{G-LP}(\omega) = e^{-0.11\left[\omega\tau^{(v)}\right]^2} e^{-i\omega\tau^{(v)}}. \tag{9.19}$$

The low-pass transfer functions of (9.18) and (9.19) have a negative phase of $-\tan^{-1}\left(\omega\tau^{(c)}\right)$ and $-\omega\tau^{(v)}$, respectively. Therefore, (9.17) shows that the phase of $\mathbf{O}_{F,\dot{O}}$ is more negative than that of $\mathbf{cbf} - \mathbf{cmro}_2$; in other words, the oscillations of oxy-hemoglobin concentration trail the driving oscillations of blood flow and metabolic rate of oxygen. This is the reason that the phasor $\mathbf{cbf}(\omega) - \mathbf{cmro}_2(\omega)$ is not aligned (i.e. in phase) with the phasor $\mathbf{O}_{F,\dot{O}}$ in Fig. 9.2, and (9.17) specifies the phase angle (indicated in Fig. 9.2) between these two phasors. This phase lag is the frequency-domain equivalent of the delay between dynamic changes in the tissue concentration of oxy-hemoglobin and the driving changes in CBF or CMRO$_2$ that were observed in the time-domain case in relation to (9.11).

9.2.4.3 Modeling Cerebral Autoregulation

Cerebral autoregulation is the homeostatic regulation of cerebral blood flow that maintains a relatively constant brain perfusion under conditions of variable mean arterial pressure. Because relatively fast blood pressure changes (on a time scale of less than one second) do not typically result in effective autoregulation, whereas slower blood pressure changes are effectively compensated by cerebrovascular resistance changes, autoregulation is often modeled with high-pass filters. Such filters are usually applied to linear systems that consider arterial blood pressure as the input and cerebral blood flow as the output. In cases where continuous arterial blood pressure measurements are not available, we have considered a high-pass filter between cerebral blood volume (the input) and cerebral blood flow (the output). Of course, cerebral blood volume is not necessarily a suitable surrogate for mean arterial pressure. However, blood volume measured by NIRS is mostly representative of the microvasculature, with greater contributions from the more compliant venous compartments, and it was reported that dynamic NIRS measurements of blood volume closely match (except from a slight delay) mean arterial pressure traces [32].

In the time domain, the relationship between blood volume and blood flow is:

$$\text{cbf}(t) = k h^{(AR)}_{RC-HP}(t) * \text{cbv}(t), \tag{9.20}$$

where k is a constant and $h^{(AR)}_{RC-HP}(t)$ is the high-pass (HP) impulse response function, approximated here by the one for a high-pass resistor-capacitor (RC) equivalent electrical circuit:

$$h^{(AR)}_{RC-HP}(t) = \delta(t) - u(t)\frac{1}{\tau^{(AR)}}e^{-t/\tau^{(AR)}}, \tag{9.21}$$

where $u(t)$ is the unit step function and $\tau^{(AR)}$ is the time constant for autoregulation.

The corresponding frequency-domain relationships, expressed in phasor notation and invoking the autoregulation transfer function, given by the Fourier transform of (9.21), are:

$$\mathbf{cbf}(\omega) = k\mathcal{H}^{(AR)}_{RC-HP}(\omega)\mathbf{cbv}(\omega), \tag{9.22}$$

$$\mathcal{H}^{(AR)}_{RC-HP}(\omega) = \frac{i\omega\tau^{(AR)}}{1 + i\omega\tau^{(AR)}}. \tag{9.23}$$

The high-pass transfer function of (9.23) has a positive phase of $\tan^{-1}(1/(\omega\tau^{(c)}))$. Therefore, (9.22) shows that the phase of **cbf** is more positive than that of **cbv**; in other words, CBF oscillations lead CBV oscillations. This is reflected in the time domain by the fact that higher frequencies (corresponding to temporal features having faster time scales) are retained by the high pass filter. Therefore, peak values of dynamic changes are reached at earlier times for cbf(t) than for cbv(t).

9.2.5 Coherent Hemodynamics Spectroscopy (CHS)

The above Sects. 9.2.1–9.2.4 provide the conceptual framework for the novel technique of coherent hemodynamics spectroscopy, or CHS [19]. CHS is based on coherent hemodynamics that allow for considering, at least to a first approximation, the microvasculature to behave as a linear time invariant system. The basic assumption is that, over the observation period, there is a high level of coherence between the driving physiological dynamics of blood volume, blood flow, and metabolic rate of oxygen, and the resulting, measurable dynamics of oxy- and deoxy-hemoglobin concentrations in tissue. Under these conditions, the above analytical treatment holds, and one can perform quantitative CHS measurements either in the time domain (considering general temporal dynamics such as those illustrated in Fig. 9.1a) or in the frequency domain (considering oscillatory dynamics such as those illustrated in Fig. 9.1b).

Quantitative CHS allows for the determination of a number of physiological quantities related to tissue perfusion, oxygenation, and hemodynamics that can find significant applications in diagnostic or functional studies of living tissue. For example, we have shown the ability of CHS to measure the cerebral capillary transit time in hemodialysis patients and healthy controls [33], the cutoff frequency for autoregulation in healthy subjects under normal breathing conditions and during hyperventilation-induced hypocapnia [34], and the absolute value and relative dynamics of cerebral blood flow [35].

9.3 Hybrid Method Based on Two-Layer Diffusion Theory and the Modified Beer-Lambert Law to Study Cerebral Hemodynamic Oscillations

As discussed in Sect. 9.1, some tissues cannot be assumed to be optically homogeneous when measured with non-invasive NIRS from the tissue surface. In particular, the human head features tissue layers such as the scalp, skull, dura mater and subdural space, arachnoid mater and subarachnoid space, grey matter and white matter. These different tissue types are characterized by different optical properties (i.e. absorption and reduced scattering coefficients) that must be taken into account to accurately describe light propagation from the illumination to the collection points. There is a vast literature related to diffusive layered models that can be classified into theoretical or phantom studies [36–39], and in vivo studies [23–25]. These studies were aimed at measuring either the absolute optical properties of layered tissue-like phantoms or the baseline (i.e. steady state) optical properties of layered biological tissues.

In addition to the study of steady state conditions, it is important to apply two or multi-layered models also for studying small perturbations in the optical properties of heterogeneous tissues such as those associated with cerebral and extra-cerebral hemodynamic changes, either spontaneous or induced. In particular, we could apply diffusive layered models to analyze CHS data and extract the relevant physiological parameters that pertain to the brain. However, due to signal-to-noise considerations, detecting small hemodynamic oscillations using time-resolved methods is challenging. For example, in frequency domain (FD) NIRS, expected perturbations in the absorption coefficient as induced by typical CHS maneuvers, should induce phase changes that are comparable to the noise level. For these reasons, we propose to take advantage of both FD and continuous wave (CW) NIRS measurements in order to measure the oscillations in the concentration of oxy- and deoxy-hemoglobin in brain tissue during typical CHS protocols.

The FD component of our method is based on FD NIRS measurements to estimate the baseline optical properties of the head, which is considered as a two-layer medium, with the first layer comprising extracerebral tissue, and the second layer representing the brain. By using an inversion procedure based on a two-layer diffusion model [25], we are able to derive not only the optical properties of the two layers but also the thickness of the extracerebral tissue layer. In our previous work, the inversion procedure was tested on both phantoms and human subjects [25]. Knowledge of the optical properties of the two layers and the thickness of the first layer allows for the estimation of the partial mean pathlengths travelled by detected photons in each of the two layers, which are required for disentangling the contributions to the measured signal from hemodynamic oscillations occurring in the extracerebral tissue from those occurring in the brain. In the next two sections we describe in detail the mathematical models of the proposed method of data analysis.

9.3.1 Diffusion Theory for a Two-Layered Cylindrical Medium

A two-layer diffusive medium is divided into a top region (first layer) and bottom region (second layer). The two media have different optical properties, namely the absorption coefficient (μ_a) and the reduced scattering coefficient (μ'_s). The diffusion equation for an intensity-modulated point source in the frequency domain (FD) is written as:

$$\nabla \cdot [D_0(\mathbf{r})\nabla \phi(\mathbf{r}, \Omega)] - \left[\mu_a(\mathbf{r}) + i\frac{\Omega}{v}\right]\phi(\mathbf{r}, \Omega) = -P_{AC}(\Omega)\delta(\mathbf{r}). \qquad (9.24)$$

In (9.24) ϕ is the fluence rate, i.e. the optical power per unit area impinging from all directions (units: W/m^2) at an arbitrary field point inside the medium (at position vector: $\mathbf{r}$), $D_0 = 1/(3\mu'_s)$ is the optical diffusion coefficient, v is the speed of light in the medium, Ω is the angular modulation frequency of the light intensity (in this study $\Omega/(2\pi)$ is 110 MHz), δ is the Dirac delta, and P_{AC} is the source power. Here we consider a cylindrical geometry, where one of the circular bases of the cylinder acts as the surface available to the optical probe (light sources and optical detectors), and the physical size of the cylinder (height and diameter) is large enough to neglect any boundary effects at the sides and at the circular base opposite to the one accessible to the optical probe. For a point source incident at the center of the circular base, the general solution of the two-layer diffusion equation in cylindrical coordinates ($\mathbf{r} = (\rho, \theta, z)$; z is the direction of the cylinder's axis pointing inside the medium) is given by [40]:

$$\phi_k(\mathbf{r}, \Omega) = \frac{P_{AC}(\Omega)}{\pi a'^2}\sum_{n=1}^{\infty} G_k(s_n, z, \Omega)J_0(s_n\rho)J_1^{-2}(a's_n) \qquad (9.25)$$

where ϕ_k is the fluence rate in the kth layer of the medium ($k = 1, 2$), s_n are the positive roots of the 0th-order Bessel function of the first kind divided by $a' = a + z_b$, (where a is the radius of the cylinder), and J_m is the Bessel function of the first kind of order m. Also, z_b is the distance between the extrapolated and the real boundary, z_b

$= 2D_{01}(1 + R_{eff})/(1 - R_{eff})$, R_{eff} is the fraction of photons that are internally diffusely reflected at the cylinder boundary and D_{01} is the optical diffusion coefficient in the first layer. Here, we report the solution for G_k only for the first layer ($k = 1$), since it is the layer where the reflectance is calculated. For the first layer (the one illuminated by the light source), G_1 is given by the following expression:

$$G_1(s_n, z, \Omega) = \frac{\exp(-\alpha_1|z - z_0|) - \exp[\alpha_1(z + z_0 + 2z_b)]}{2D_{01}\alpha_1}$$
$$+ \frac{\sinh[\alpha_1(z_0 + z_b)]\sinh[\alpha_1(z + z_b)]}{D_{01}\alpha_1 \exp[\alpha_1(L + z_b)]}$$
$$\times \frac{D_{01}\alpha_1 - D_{02}\alpha_2}{D_{01}\alpha_1 \cosh[\alpha_1(L + z_b)] + D_{02}\alpha_2 \sinh[\alpha_1(L + z_b)]} \qquad (9.26)$$

where L is the thickness of the first layer, z_0 is the mean transport scattering length in the top layer ($1/\mu'_{s1}$), and α_k ($k = 1, 2$) is given by:

$$\alpha_k = \sqrt{\frac{\mu_{ak}}{D_{0k}} + s_n^2 + \frac{i\Omega}{D_{0k}v}} \qquad (9.27)$$

Equation (9.26) is attained in the limiting case of an infinite second layer in the z direction, and for the situation where the two layers have the same refractive index [40]. In (9.27), D_{0k} and μ_{ak} are the kth layer diffusion and the absorption coefficients. For the calculation of the optical diffuse reflectance (R), one may apply Fick's law:

$$R(\rho, \Omega) = D_{01}\frac{\partial}{\partial z}\phi_1(\rho, z, \Omega)\Big|_{z=0} \qquad (9.28)$$

From the reflectance expression (as fully described by a complex function), one can determine the amplitude, or AC(ρ, Ω) and the phase $\theta(\rho, \Omega)$, the two main quantities measured in FD NIRS, as follows: AC(ρ, Ω) = |$R(\rho, \Omega)$|, $\theta(\rho, \Omega) = Arg[R(\rho, \Omega)]$.

The above solution of the diffusion equation was used as the forward solver in an inversion methodology, based on the Levenberg-Marquardt method which allows to recover the two layers optical properties (μ_a, μ'_s) and the first layer thickness from measured AC amplitude and phase data at six source detector separations [25]. When data taken on human subjects are considered, the inversion procedure is run by using average values of AC amplitude and phase data acquired during baseline, allowing us to recover the average baseline oxy- and deoxy-hemoglobin concentrations in extracerebral tissue and in the brain.

9.3.2 mBLL for Measuring Hemodynamic Oscillations

CW intensity oscillations due to absorption oscillations in tissue are given (for the case of a two-layered model of tissue) by the phasor equation:

$$\mathbf{i}(d_j, \lambda, \omega) = -\sum_{i=1}^{2} \langle l_i(d_j, \lambda) \rangle \mathbf{\mu}_{ai}(\lambda, \omega) \tag{9.29}$$

In (9.29), $\mathbf{i}(d_j, \lambda, \omega)$ is the phasor associated with sinusoidal intensity oscillations normalized to the baseline, or average intensity: $\frac{I(d_j, \lambda, \omega, t) - I_b(d_j, \lambda)}{I_b(d_j, \lambda)}$, where $I(d_j, \lambda, \omega, t)$ and $I_b(d_j, \lambda)$ are the instantaneous and baseline (or average) intensities, respectively, d_j is the source detector separation ($j = 1, \ldots, 6$ in this work), λ is the wavelength ($\lambda = 690, 830$ nm in this work) and i is the layer index ($i = 1, 2$). $\langle l_i(d_j, \lambda) \rangle$ is the mean optical pathlength in layer i for photons detected at d_j. $\mathbf{\mu}_{ai}(\lambda, \omega)$ is the phasor associated with sinusoidal absorption oscillations in layer i, normalized to the baseline, or average absorption coefficient: $\frac{\mu_{ai}(\lambda, \omega, t) - \mu_{aib}(\lambda)}{\mu_{aib}(\lambda)}$, where $\mu_{ai}(\lambda, \omega, t)$, and $\mu_{aib}(\lambda)$ are the instantaneous and baseline (or average) absorption coefficients, respectively, in layer i. Note that $\mu_{aib}(\lambda)$ is measured by the FD method described in the previous section for both layers $i = 1, 2$. The phasors associated with oxy- and deoxy-hemoglobin oscillations in the two layers (which comprise contributions from both blood volume and blood flow oscillations) are related to the absorption phasors by the relationship:

$$\mathbf{\mu}_{ai}(\lambda, \omega) = \varepsilon_{HbO2}(\lambda)\mathbf{O}_i(\omega) + \varepsilon_{Hb}(\lambda)\mathbf{D}_i(\omega) \tag{9.30}$$

where $\mathbf{O}_i$ and $\mathbf{D}_i$ are the oxy- and deoxyhemoglobin phasors, respectively, in layer i, and ε_{HbO2} and ε_{Hb} are the wavelength-dependent extinction coefficients of oxy- and deoxy-hemoglobin.

Equation (9.29) represents a generalized mBLL for a two-layered medium, and it can be re-written in the form of the standard mBLL by introducing an effective absorption phasor that depends on the source-detector separation:

$$\mathbf{i}(d_j, \lambda, \omega) = -\langle L(d_j, \lambda) \rangle \mathbf{\mu}_{a,\text{eff}}(d_j, \lambda, \omega) \tag{9.31}$$

In (9.31), the intensity phasor is expressed in terms of an effective absorption coefficient phasor $\mathbf{\mu}_{a,\text{eff}}(d_j, \lambda, \omega)$ (which depends on d_j) and the total mean pathlength, $\langle L(d_j, \lambda) \rangle$, traveled by the photons that are detected at a source-detector distance d_j. The effective absorption coefficient phasor is related to effective oxy- and deoxy-hemoglobin concentration phasors by a relationship formally similar to (9.30):

$$\mathbf{\mu}_{a,\text{eff}}(d_j, \lambda, \omega) = \varepsilon_{HbO2}(\lambda)\mathbf{O}_{\text{eff}}(d_j, \omega) + \varepsilon_{Hb}(\lambda)\mathbf{D}_{\text{eff}}(d_j, \omega) \tag{9.32}$$

By using (9.29)–(9.32), we derive the following system of equations:

$$\mathbf{O}_{\text{eff}}(d_j, \omega) = \alpha_1(d_j)\mathbf{D}_1(\omega) + \alpha_2(d_j)\mathbf{O}_1 + \alpha_3(d_j)\mathbf{D}_2(\omega) + \alpha_4(d_j)\mathbf{O}_2 \qquad (9.33)$$

$$\mathbf{D}_{\text{eff}}(d_j, \omega) = \beta_1(d_j)\mathbf{D}_1(\omega) + \beta_2(d_j)\mathbf{O}_1 + \beta_3(d_j)\mathbf{D}_2(\omega) + \beta_4(d_j)\mathbf{O}_2 \qquad (9.34)$$

where the coefficients α_i and β_i depend on both wavelengths, through the extinction coefficients and also through the ratio of partial to total mean optical pathlengths $\left(\frac{\langle l_i(d_j, \lambda)\rangle}{\langle L(d_j, \lambda)\rangle}\right)$ [30]. The coefficients α_1, α_3 (in (9.33) for oxy-hemoglobin) and β_2, β_4 (in (9.34) for deoxy-hemoglobin) are called cross talk coefficients, and are much smaller than the other coefficients, and usually could be omitted without significantly affecting the results. However, in this study we used (9.33) and (9.34) with no approximations.

We observe that knowledge of the baseline values of the optical properties in the two layers and the thickness of the first layer allows us to calculate all of the coefficients α and β in (9.33) and (9.34). In this work, the shortest source-detector separation was $d_1 = 8$ mm. Considering that the combined scalp and skull thickness in adult human is typically 8–16 mm, we made the reasonable assumption that $\mathbf{O}_{\text{eff}}(d_1, \omega) = \mathbf{O}_1(\omega)$ and $\mathbf{D}_{\text{eff}}(d_1, \omega) = \mathbf{D}_1(\omega)$; in other words, the oxy- and deoxy-hemoglobin phasors obtained with the mBLL from the measured optical signals at the shortest source-detector separation are taken to coincide with the oxy- and deoxyhemoglobin phasors of the top layer. Then, we used (9.33) and (9.34) to retrieve the phasors of oxy- and deoxy-hemoglobin concentrations in the bottom layer, $\mathbf{O}_2(\omega)$ and $\mathbf{D}_2(\omega)$, by using the oxy- and deoxy-hemoglobin phasors obtained with the mBLL from the measured optical signals at the longest source-detector separation, $\mathbf{O}_{\text{eff}}(d_6, \omega)$ and $\mathbf{D}_{\text{eff}}(d_6, \omega)$. The reason for using the longest source-detector distance is that, at this distance, the measured optical intensities are most sensitive to the bottom, cerebral tissue layer.

9.4   Experimental Methods

A commercial frequency-domain tissue spectrometer (OxiplexTS, ISS Inc., Champaign, IL) was used in the NIRS measurements that were performed. A frequency of 110 MHz was used for modulating the laser intensity outputs. An optical probe was connected to the spectrometer through optical fibers distributing 690 and 830 nm wavelength light at six different locations, separated 8, 13, 18, 28, 33, and 38 mm from a single collection optical fiber. The optical probe was positioned against the subject's forehead right side and fixed with a flexible headband. The instrument was calibrated through an optical phantom of known optical properties, by means of the so called multidistance FD NIRS procedure [41].

Pneumatic thigh cuffs were swaddled around the person's thighs and plugged to an automated cuff inflation system (E-20 Rapid Cuff Inflation System, D. E. Hokanson, Inc., Bellevue, WA). The thigh cuffs air pressure was continuously monitored

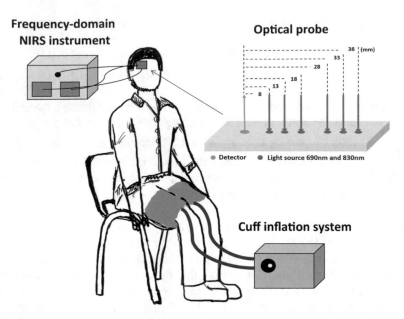

Fig. 9.3 Schematic of the experimental setup, including the frequency-domain NIRS instrument, the pneumatic thigh cuff, and the automated cuff inflation system. A detail of the optical probe is shown to specify the six source-detector distances used

through a digital manometer (Series 626 Pressure Transmitter, Dwyer Instruments, Inc., Michigan City, IN). The thigh cuff pressure monitor analogue outputs were fed to auxiliary inputs of the NIRS instrument for concomitant recordings together with the NIRS data. A 25 year old healthy female was subjected to an eight cycles protocol, where in each cycle the thigh cuff was inflated to a pressure of 190 mmHg for 9 s and released for 8 s ($f = 0.059$ Hz, where $f = \omega/2\pi$ is the frequency of the cuff inflation/deflation cycles).

The analytic signal method [42, 43] was used to associate phasors (i.e. an amplitude and phase) to the oscillations of oxy- and deoxy-hemoglobin concentrations, which were obtained by using the mBLL as explained in Sect. 9.3.2 ((9.31) and (9.32)). The overall experimental setup layout is displayed in Fig. 9.3.

9.5   Results

The frequency-domain multi-distance NIRS data collected in vivo, analyzed with the two-layer diffusion model of Sect. 9.3.1, yielded the values of the baseline optical properties of the two tissue layers and of the thickness of the top layer that are reported in Table 9.1. On the basis of these baseline optical properties and thickness of top layer, we have computed the α and β coefficients in the system of (9.33) and (9.34),

Table 9.1 Baseline optical properties and top layer thickness obtained on the head of a human subject with multi-distance FD NIRS data and a two-layer diffusion model

		λ_1 (690 nm)	λ_2 (830 nm)
Top layer	μ_{a1} (cm^{-1})	0.072	0.068
	μ_{s1}' (cm^{-1})	12.4	12.5
	Thickness (mm)	9	
Bottom layer	μ_{a2} (cm^{-1})	0.089	0.10
	μ_{s2}' (cm^{-1})	2.7	2.1

Table 9.2 Coefficients of the linear combinations of the oxy- and deoxy-hemoglobin concentration phasors in the top and bottom tissue layer ($\mathbf{O}_1$, $\mathbf{D}_1$, $\mathbf{O}_2$, $\mathbf{D}_2$) that result in the effective phasors obtained from CW NIRS data at source-detector separation $d_6 = 38$ mm with the modified Beer-Lambert law ($\mathbf{O}_{\text{eff}}(d_6)$, $\mathbf{D}_{\text{eff}}(d_6)$) (see (9.33) and (9.34))

$\alpha_1(d_6)$	0.062
$\alpha_2(d_6)$	0.63
$\alpha_3(d_6)$	−0.062
$\alpha_4(d_6)$	0.37
$\beta_1(d_6)$	0.54
$\beta_2(d_6)$	−0.016
$\beta_3(d_6)$	0.46
$\beta_4(d_6)$	0.016

for the greater source-detector distance ($d_6 = 38$ mm), as reported in Table 9.2. This is the first step of our hybrid approach: the determination of the coefficients of the linear system of (9.33) and (9.34) that allows us to translate single-distance, CW NIRS measurements of effective oxy- and deoxy-hemoglobin concentration phasors at short distance ($\mathbf{O}_{\text{eff}}$, $\mathbf{D}_{\text{eff}}$ at $d_1 = 8$ mm) and long distance ($\mathbf{O}_{\text{eff}}$, $\mathbf{D}_{\text{eff}}$ at $d_6 = 38$ mm) into the actual hemoglobin concentration phasors in the top layer ($\mathbf{O}_1$, $\mathbf{D}_1$) and bottom layer ($\mathbf{O}_2$, $\mathbf{D}_2$) of the investigated tissue.

Figure 9.4 shows the phasors of the concentrations of oxy- and deoxy-hemoglobin at six source-detector separations derived with the mBLL (top phasors in Fig. 9.4). Even though we have used only the effective phasors measured at the first and the sixth source-detector distances for our method of recovering $\mathbf{O}_2(\omega)$ and $\mathbf{D}_2(\omega)$, in Fig. 9.4 we also show the other effective phasors to illustrate that both oxy- and deoxy-hemoglobin effective phasors rotate clockwise as they are measured farther and farther from the detector. Because larger source-detector separations feature enhanced sensitivities to deeper tissues, this trend is indicative of different oscillations occurring in the extracerebral tissue and in the brain. Specifically, these results indicate that hemodynamic oscillations in the bottom layer have a lower phase than those in the top layer. In other words, deeper hemodynamic oscillations lag superficial hemodynamic oscillations.

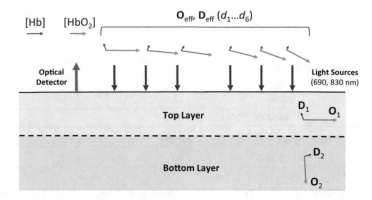

Fig. 9.4 Schematic illustration of the two-layer tissue model and the set of illumination and collection optical fibers at six source-detector distances (8, 13, 18, 28, 33, and 38 mm). The set of phasors above each illumination optical fiber are the effective oxy- and deoxy-hemoglobin concentration phasors obtained from single-distance CW NIRS data analyzed with the modified Beer-Lambert law. The phasors inside the top and bottom layers represent the actual hemodynamic oscillations in the two layers, as obtained by our hybrid FD/CW NIRS approach

The phases of all phasors were taken relative to the phase of $\mathbf{O}_{\mathrm{eff}}(d_1, \omega) = \mathbf{O}_1(\omega)$ (we recall that the oxy-hemoglobin phasor obtained with the mBLL from the optical data at the shortest source-detector distance, d_1, was taken to coincide with the oxy-hemoglobin phasor of the top layer), so that, by definition, Arg $(\mathbf{O}_1(\omega)) = 0°$. With this convention, the effective oxy- and deoxy-hemoglobin phasors obtained from CW NIRS data and the mBLL at the shortest (d_1) and longest (d_6) source-detector distances were: $\mathbf{O}_{\mathrm{eff}}(d_1) = 0.67\,\mu\mathrm{M}\,\angle 0°$; $\mathbf{D}_{\mathrm{eff}}(d_1) = 0.096\,\mu\mathrm{M}\,\angle 103°$; $\mathbf{O}_{\mathrm{eff}}(d_6) = 0.49\,\mu\mathrm{M}\,\angle -28°$; $\mathbf{D}_{\mathrm{eff}}(d_6) = 0.067\,\mu\mathrm{M}\,\angle 55°$.

As described in Sect. 9.3.2, we identified $\mathbf{O}_{\mathrm{eff}}(d_1)$ and $\mathbf{D}_{\mathrm{eff}}(d_1)$ with the oxy- and deoxy-hemoglobin concentration phasors in the top layer: $\mathbf{O}_1 = 0.67\,\mu\mathrm{M}\,\angle 0°$ and $\mathbf{D}_1 = 0.096\,\mu\mathrm{M}\,\angle 103°$. From (9.33) and (9.34), using the coefficients of Table 9.2, we obtained the oxy- and deoxy-hemoglobin concentration phasors in the bottom layer: $\mathbf{O}_2 = 0.62\,\mu\mathrm{M}\,\angle -87°$ and $\mathbf{D}_2 = 0.11\,\mu\mathrm{M}\,\angle 13°$. These phasor results for the top and bottom tissue layers, which reflect hemodynamic oscillations at the induced frequency of 0.059 Hz, confirm the qualitative observation that hemodynamic oscillations in the bottom layer lag those in the top layer (in this case, by about 90°).

9.6   Discussion and Conclusions

We have described a novel technique, CHS, to perform quantitative studies of cerebral hemodynamics on the basis of data collected with NIRS or fMRI. This technique is combined with a dedicated analytical model that describes arbitrary time courses of oxy- and deoxy-hemoglobin concentrations in tissue (time-domain representation), or oscillatory hemodynamics at well-defined frequencies (frequency-domain representation). This analytical model enables the translation of measurable hemoglobin concentrations in tissue into physiological quantities of interest for diagnostic assessment, functional studies, or characterization of tissue perfusion. The dynamic study of blood perfusion and oxygenation has been performed for many years using both NIRS and fMRI in a number of research areas. For example, spontaneous low-frequency hemodynamic oscillations have been investigated in studies of resting state functional connectivity based on fMRI [44–46] or NIRS [47–50]. Even more closely related to CHS, several NIRS studies have investigated the relative phase of cerebral [Hb] and [HbO$_2$] oscillations occurring spontaneously [51] or in response to paced breathing [32, 52, 53]. The innovative aspects of CHS are the systematic nature of the measurements of coherent hemodynamics, and the quantitative analysis afforded by the analytical mathematical model.

In the case of non-invasive NIRS studies of the human brain, one has to take into consideration the presence of superficial, non-cerebral tissue layers that may confound optical measurements of the brain cortex. This is a well-known limiting factor of non-invasive functional NIRS, which has been tackled in a number of ways, typically based on collecting data at multiple source-detector separations in combination with regression methods [26], spectral analysis [54], independent component analysis [55], or the computation of partial mean pathlengths in two-layered media [56].

This chapter describes a hybrid FD/CW method based on the mBLL and on a two-layer diffusion model to disentangle the contributions to the optical signals from induced hemodynamic oscillations occurring in the extracerebral (superficial) tissue and in the (deeper) brain tissue. This problem was tackled by using a two-layer diffusion model and the combination of NIRS data collected at multiple source-detector separations. This two-layer diffusion model has by now been used in the evaluation of baseline optical features and hemoglobin concentrations in two "effective" head tissue layers. Frequently, in NIRS, the dynamics of hemoglobin species are calculated by means of the modified Beer-Lambert law (mBLL), which considers homogeneous absorption changes in a light probed tissue. As this premise can be violated under a variety of conditions, it would be fundamental to distinguish the hemodynamic oscillations taking place in the extra-cerebral layers, namely scalp and skull, from those taking place in the brain. Thereupon a more realistic two distinct layers model of the head has been considered. The diffusion equation solution in the FD corresponding to a two-layer configuration was used to recover the absorption baseline values, the reduced scattering coefficients of both layers, and also the first layer thickness, from frequency-domain data at six source-detector separations. These baseline values were used to calculate the partial (in each layer) and total (in the entire medium) mean optical pathlengths which allowed us to recover the oxy- and deoxy-hemoglobin hemodynamic oscillations in the two tissue layers by using CW data. This is the first step to recover relevant physiological parameters specific to the brain by using CHS.

In summary, we have described CHS, a new technique to investigate cerebral hemodynamics, and a first attempt towards the creation of depth-resolved CHS.

The ability to perform accurate measurements of the cerebral hemodynamics with non-invasive optical techniques can ultimately result in a powerful tool towards the non-invasive assessment of cerebral perfusion, autoregulation, functional activation, and overall tissue viability in vivo.

Acknowledgements This research is supported by the US National Institutes of Health (Grants no. R01-NS095334 and R21-EB020347).

References

1. R.B. Buxton, L.R. Frank, A model of the coupling between cerebral blood flow and oxygen metabolism during neural stimulation. J. Cereb. Blood Flow Metab. **17**, 64–72 (1997)
2. J.B. Mandeville, J.J.A. Marota, C. Ayata, G. Zaharchuk, M.A. Moskowitz, B.R. Rosen, R.M. Weisskoff, Evidence of a cerebrovascular postarteriole windkessel with delayed compliance. J. Cereb. Blood Flow Metab. **19**, 679–689 (1999)
3. M. Ferrari, V. Quaresima, A brief review on the history of human functional near-infrared spectroscopy (fNIRS) development and fields of application. NeuroImage **63**, 921–935 (2012)
4. F. Scholkmann, S. Kleiser, A.J. Metz, R. Zimmermann, J.M. Pavia, U. Wolf, M. Wolf, A review on continuous wave functional near-infrared spectroscopy and imaging instrumentation and methodology. NeuroImage **85**, 6–27 (2014)
5. A. Torricelli, D. Contini, A. Pifferi, M. Caffini, R. Re, L. Zucchelli, L. Spinelli, Time domain functional NIRS imaging for human brain mapping. NeuroImage **85**, 28–50 (2014)
6. M. Smith, Shedding light on the adult brain: a review of the clinical applications of near-infrared spectroscopy. Philos. Trans. R. Soc. A **369**, 4452–4469 (2011)
7. A. Casati, G. Fanelli, P. Pietropaoli, R. Tufano, G. Danelli, G. Fierro, G. De Cosmo, G. Servillo, Continuous monitoring of cerebral oxygen saturation in elderly patients undergoing major abdominal surgery minimizes brain exposure to potential hypoxia. Anesth. Analg. **101**, 740–747 (2005)
8. M.M. Tisdall, I. Tachtsidis, T.S. Leung, C.E. Elwell, M. Smith, Increase in cerebral aerobic metabolism by normobaric hyperoxia after traumatic brain injury. J. Neurosurg. **109**, 424–432 (2008)
9. S.R. Leal-Noval, A. Cayuela, V. Arellano-Orden, A. Marin-Caraballos, V. Padilla, C. Ferrándiz-Millón, Y. Corcia, C. García-Alfaro, R. Amaya-Villar, F. Murillo-Cabezas, Invasive and noninvasive assessment of cerebral oxygenation in patients with severe traumatic brain injury. Intens. Care Med. **36**, 1309–1317 (2010)
10. F. Fabbri, M.E. Henry, P.F. Renshaw, S. Nadgir, B.L. Ehrenberg, M.A. Franceschini, S. Fantini, Bilateral near-infrared monitoring of the cerebral concentration and oxygen saturation of hemoglobin during right unilateral electro-convulsive therapy. Brain Res. **992**, 193–204 (2003)
11. V. Toronov, A. Webb, J.H. Choi, M. Wolf, L. Safonova, U. Wolf, E. Gratton, Study of local cerebral hemodynamics by frequency-domain near-infrared spectroscopy and correlation with simultaneously acquired functional magnetic resonance imaging. Opt. Express **9**, 417–427 (2001)
12. A.M. Siegel, J.P. Culver, J.B. Mandeville, D.A. Boas, Temporal comparison of functional brain imaging with diffuse optical tomography and fMRI during rat forepaw stimulation. Phys. Med. Biol. **48**, 1391–1403 (2003)
13. A. Sassaroli, B. deB. Frederick, Y. Tong, P.F. Renshaw, S. Fantini, Spatially weighted BOLD signal for comparison of functional magnetic resonance imaging and near-infrared imaging of the brain. NeuroImage **33**, 505–514 (2006)
14. X. Cui, S. Bray, D.M. Bryant, G.H. Glover, A.L. Reiss, A quantitative comparison of NIRS and fMRI across multiple cognitive tasks. NeuroImage **54**, 2808–2821 (2011)

15. T.J. Huppert, M.S. Allen, H. Benav, P.B. Jones, D.A. Boas, A multicompartment vascular model for inferring baseline and functional changes in cerebral oxygen metabolism and arterial dilation. J. Cereb. Blood Flow Metab. **27**, 1262–1279 (2007)
16. M. Banaji, A. Mallet, C.E. Elwell, P. Nicholls, C.E. Cooper, A model of brain circulation and metabolism: NIRS signal changes during physiological challenges. PLoS Comput. Biol. **4**, e1000212 (2008)
17. D.A. Boas, S.R. Jones, A. Devor, T.J. Huppert, A.M. Dale, A vascular anatomical network model of the spatio-temporal response to brain activation. NeuroImage **40**, 1116–1129 (2008)
18. S.G. Diamond, K.L. Perdue, D.A. Boas, A cerebrovascular response model for functional neuroimaging including cerebral autoregulation. Math. Biosci. **220**, 102–117 (2009)
19. S. Fantini, Dynamic model for the tissue concentration and oxygen saturation of hemoglobin in relation to blood volume, flow velocity, and oxygen consumption: implications for functional neuroimaging and coherent hemodynamics spectroscopy (CHS). NeuroImage **85**, 202–221 (2014)
20. S. Fantini, A new hemodynamic model shows that temporal perturbations of cerebral blood flow and metabolic rate of oxygen cannot be measured individually using functional near-infrared spectroscopy. Physiol. Meas. **35**, N1–N9 (2014)
21. D.T. Delpy, M. Cope, P. van der Zee, S. Arridge, S. Wray, J. Wyatt, Estimation of optical path length through tissue from direct time of flight measurements. Phys. Med. Biol. **33**, 1433–1442 (1988)
22. A. Sassaroli, S. Fantini, Comment on the modified Beer-Lambert law for scattering media. Phys. Med. Biol. **49**, N1–N3 (2004)
23. J. Choi, M. Wolf, V. Toronov, U. Wolf, C. Polzonetti, D. Hueber, L.P. Safonova, R. Gupta, A. Michalos, W. Mantulin, E. Gratton, Noninvasive determination of the optical properties of adult brain: near-infrared spectroscopy approach. J. Biomed. Opt. **9**, 221–229 (2004)
24. L. Gagnon, C. Gauthier, R.D. Hoge, F. Lesage, J. Selb, D.A. Boas, Double-layer estimation of intra- and extracerebral hemoglobin concentration with a time-resolved system. J. Biomed. Opt. **13**, 054019 (2008)
25. B. Hallacoglu, A. Sassaroli, S. Fantini, Optical characterization of two-layered turbid media for non-invasive, absolute oximetry in cerebral and extracerebral tissue. PLoS ONE **8**, e64095 (2013)
26. R.B. Saager, A.J. Berger, Direct characterization and removal of interfering absorption trends in two-layer turbid media. J. Opt. Soc. Am. A **22**, 1874–1882 (2005)
27. Q. Zhang, G.E. Strangman, G. Ganis, Adaptive filtering to reduce global interference in non-invasive NIRS measures of brain activation: how well and when does it work? NeuroImage **45**, 788–794 (2009)
28. L. Gagnon, K. Perdue, D.N. Greve, D. Goldenholz, G. Kaskhedikar, D.A. Boas, Improved recovery of the hemodynamic response in diffuse optical imaging using short optode separations and state-space modeling. NeuroImage **56**, 1362–1371 (2011)
29. F. Scarpa, S. Brigadoi, S. Cutini, P. Scatturin, M. Zorzi, R. Dell'Acqua, G. Sparacino, A reference-channel based methodology to improve estimation of event-related hemodynamic response from fNIRS measurements. NeuroImage **72**, 106–119 (2013)
30. F. Zheng, A. Sassaroli, S. Fantini, Phasor representation of oxy- and deoxyhemoglobin concentrations: what is the meaning of out-of-phase oscillations as measured by near-infrared spectroscopy? J. Biomed. Opt. **15**, 040512 (2010)
31. J.M. Kainerstorfer, A. Sassaroli, S. Fantini, Optical oximetry of volume-oscillating vascular compartments: contributions from oscillatory blood flow. J. Biomed. Opt. **21**, 101408 (2016)
32. M. Reinhard, E. Wehrle-Wieland, D. Grabiak, M. Roth, B. Guschlbauer, J. Timmer, C. Weiller, A. Hetzel, Oscillatory cerebral hemodynamics: the macro- vs. microvascular level. J. Neurol. Sci. **250**, 103–109 (2006)
33. M.L. Pierro, J.M. Kainerstorfer, A. Civiletto, D.E. Wiener, A. Sassaroli, B. Hallacoglu, S. Fantini, Reduced speed of microvascular blood flow in hemodialysis patients versus healthy controls: a coherent hemodynamics spectroscopy study. J. Biomed. Opt. **19**, 026005 (2014)

34. J.M. Kainerstorfer, A. Sassaroli, K.T. Tgavalekos, S. Fantini, Cerebral autoregulation in the microvasculature measured with near-infrared spectroscopy. J. Cereb. Blood Flow Metab. **35**, 959–966 (2015)
35. S. Fantini, A. Sassaroli, J.M. Kainerstorfer, K.T. Tgavalekos, X. Zang, Non-invasive assessment of cerebral microcirculation with diffuse optics and coherent hemodynamics spectroscopy. Proc. SPIE **9690**, 96900S (2016)
36. G. Alexandrakis, D.R. Busch, G.W. Faris, M.S. Patterson, Determination of the optical properties of two-layer turbid media by use of a frequency-domain hybrid Monte Carlo diffusion model. Appl. Opt. **40**, 3810–3821 (2001)
37. T.H. Pham, T. Spott, L.O. Svaasand, B.J. Tromberg, Quantifying the properties of two layer turbid media with frequency-domain diffuse reflectance. Appl. Opt. **39**, 4733–4745 (2000)
38. F. Martelli, A. Sassaroli, S. Del Bianco, Y. Yamada, G. Zaccanti, Solution of the time-dependent diffusion equation for layered diffusive media by the eigenfunction method. Phys. Rev. E **67**, 056623 (2003)
39. J. Ripoll, V. Ntziachristos, J.P. Culver, D.N. Pattanayak, A.G. Yodh, M. Nieto-Vesperinas, Recovery of optical parameters in multiple-layered diffusive media: theory and experiments. J. Opt. Soc. Am. A **18**, 821–830 (2001)
40. A. Liemert, A. Kienle, Light diffusion in N-layered turbid media: frequency and time domains. J. Biomed. Opt. **15**, 025002 (2010)
41. S. Fantini, M.A. Franceschini, E. Gratton, Semi-infinite-geometry boundary problem for light migration in highly scattering media: a frequency-domain study in the diffusion approximation. J. Opt. Soc. Am. B **11**, 2128–2138 (1994)
42. B. Boashash, Estimating and interpreting the instantaneous frequency of a signal. Part 1: Fundamentals. Proc. IEEE **80**, 520–538 (1992)
43. M. Pierro, A. Sassaroli, P.R. Bergethon, B.L. Ehrenberg, S. Fantini, Phase-amplitude investigation of spontaneous low-frequency oscillations of cerebral hemodynamics with near-infrared spectroscopy: a sleep study in human subjects. NeuroImage **63**, 1571–1584 (2012)
44. B.B. Biswal, F.Z. Yetkin, V.M. Haughton, J.S. Hyde, Functional connectivity in the motor cortex of resting human brain using echo-planar MRI. Magn. Reson. Med. **34**, 537–541 (1995)
45. M.D. Greicius, B. Krasnow, A.L. Reiss, V. Menon, Functional connectivity in the resting brain: a network analysis of the default mode hypothesis. Proc. Nat. Acad. Sci. USA **100**, 253–258 (2003)
46. M.D. Fox, M.E. Raichle, Spontaneous fluctuations in brain activity observed with functional magnetic resonance imaging. Nature **8**, 700–711 (2007)
47. B.R. White, A.Z. Snyder, A.L. Cohen, S.E. Petersen, M.E. Raichle, B.L. Schlaggar, J.P. Culver, Resting-state functional connectivity in the human brain revealed with diffuse optical tomography. NeuroImage **47**, 148–156 (2009)
48. C.M. Lu, Y.J. Zhang, B.B. Biswal, Y.F. Zang, D.L. Peng, C.Z. Zhu, Use of fNIRS to assess resting state functional connectivity. J. Neurosci. Methods **186**, 242–249 (2010)
49. S. Sasai, H. Fumitaka, H. Watanabe, G. Taga, Frequency-specific functional connectivity in the brain during resting state revealed by NIRS. NeuroImage **56**, 252–257 (2011)
50. A. Sassaroli, M. Pierro, P.R. Bergethon, S. Fantini, Low-frequency spontaneous oscillations of cerebral hemodynamics investigated with near-infrared spectroscopy: a review. IEEE J. Sel. Top. Quantum Electron. **18**, 1478–1492 (2012)
51. G. Taga, Y. Konishi, A. Maki, T. Tachibana, M. Fujiwara, H. Koizumi, Spontaneous Oscillation of oxy- and deoxy-hemoglobin changes with a phase difference throughout the occipital cortex of newborn infants observed using non-invasive optical topography. Neurosci. Lett. **282**, 101–104 (2000)
52. H. Obrig, M. Neufang, R. Wenzel, M. Kohl, J. Steinbrink, K. Einhäupl, A. Villringer, Spontaneous low frequency oscillations of cerebral hemodynamics and metabolism in human adults. NeuroImage **12**, 623–639 (2000)
53. F. Tian, H. Niu, B. Khan, G. Alexandrakis, K. Behbehani, H. Liu, Enhanced functional brain imaging by using adaptive filtering and a depth compensation algorithm in diffuse optical tomography. IEEE Trans. Med. Imag. **30**, 1239–1251 (2011)

54. O. Pucci, V. Toronov, K. St.Lawrence, Measurement of the optical properties of a two-layer model of the human head using broadband near-infrared spectroscopy. Appl. Opt. **49**, 6324–6332 (2010)
55. T. Funane, H. Atsumori, T. Katura, A.N. Obata, H. Sato, Y. Tanikawa, E. Okada, M. Kiguchi, Quantitative evaluation of deep and shallow tissue layers' contribution to fNIRS signal using multi-distance optodes and independent component analysis. NeuroImage **85**, 150–165 (2014)
56. L. Zucchelli, D. Contini, R. Re, A. Torricelli, L. Spinelli, Method for the discrimination of superficial and deep absorption variations by time domain fNIRS. Biomed. Opt. Express **4**, 2893–2910 (2013)

Chapter 10
Study of Endogenous Fluorescence as a Function of Tissues' Conservation Using Spectral and Lifetime Measurements on Tumor or Epileptic Cortex Excision

F. Poulon, M. Zanello, A. Ibrahim, P. Varlet, B. Devaux and D. Abi Haidar

Abstract Until today the endogenous fluorescence of tissue were neglected and often consider as a source of noise in medical imaging, however recent work and future technologies seems to reconsider it as a new imaging modality in medical devices. One of the precursor fields for the use of autofluorescence in tissue is the study of cancerology, which was recognized as a powerful tool for the future of medical devices. Although many studies have been started and done in this field, there are still numerous aspects of the signal that are not well known yet such as time dependence after extraction of fresh tissues. In this work, freshly resected human samples were exanimated in order to investigate their autofluorescence changes with time. Primary results of this examination prove that fluorescence intensity and lifetime values of healthy and tumoral samples decreased slightly with time.

F. Poulon · M. Zanello · A. Ibrahim · D. Abi Haidar (✉)
IMNC Laboratory, UMR 8165, CNRS/IN2P3, Paris-Saclay University,
91405 Orsay, France
e-mail: abihaidar@imnc.in2p3.fr

M. Zanello · B. Devaux
Department of Neurosurgery, Sainte Anne Hospital, Paris, France

P. Varlet
Department of Neuropathology, Sainte Anne Hospital, Paris, France

D. Abi Haidar
Paris Diderot University, Sorbonne Paris Cité, 75013 Paris, France

© Springer Nature Switzerland AG 2018   209
P. A. Ribeiro and M. Raposo (eds.), *Optics, Photonics and Laser Technology*, Springer
Series in Optical Sciences 218, https://doi.org/10.1007/978-3-319-98548-0_10

10.1 Introduction

Neurosurgery has benefited greatly over the last years from technologies, it became a self-standing specialty during the first decades of the 20th century [1]. Since the pioneering work of Harvey Cushing, neurosurgeons worked together with scientists to improve surgical interventions for many years. Due to these collaborations, operative microscope is employed since the 1970s [2], endoscope since the 1980s [3] and echography is now routinely performed during interventions [4].

The two major primary brain tumors are the glioma (16%) and the meningioma (36%), [5], and looking at the over all of primary and second brain tumors, the metastases represents 55% of brain tumors. These three histological types are very challenging tumors to resect as they are presenting infiltration with low concentration tumors cells around the solid tumor. The actual techniques used to diagnose a tumor are the Risk of malignancy Index (RMI) and the Computed Tomography (CT) scan, both technologies cannot give a resolution under millimeter scale, they are also difficult to bring to the operating room and consequently cannot give information in real time on the environment of the tumors. Nowadays there are no techniques to identify these regions, which appear healthy to the naked eye.

Actual philosophy used by the surgeon is a maximum extent of the resection while preserving the main brain function in the surrounding areas [6, 7]. However, it is still impossible for a neurosurgeon to know in real time and with assurance if all the margins around the solid tumor were correctly cleared due to the lack of contrast in this region. Only the examination of biopsies, after surgery, is in fact the right way to really know. Nevertheless, the results of the biopsy are obtained only few days later, and if it appears that the surgeon did not took all the tumor, a second surgery is required, with added costs and bringing discomfort to the patient. In order to conciliate an efficient surgery and a high post-operative quality of life, developing new optical tools with higher resolution and compatible to the surgery room appears as a necessity. This limit can be overcome by modern optical imaging based on fluorescence contrast. Potential optical devices with a micrometric resolution are currently explored in neurosurgery and neurooncology [8, 9]. They are most of the time associated with the use of external markers such as 5-ALA, indocyanine or fluorescein sodium, introducing a potential bias. If fluorescence analysis is a young and promising technique, the microscopic probes under in development today are mainly confocal microscopes deported by optical fiber. These systems present some major drawbacks: low level of beam penetration, strong photodamage, and also having only one contrast to look at the data, which is difficult to exploit and to extend to a large cohort to give a definite response on the nature of tissues and cells. Future developments on exogenous or endogenous fluorescence are all going into a multimodality way to improve accuracy of the analysis [10].

To address this key challenge in neurosurgery, one proposes the development of a miniaturized multimodal nonlinear endomicroscope. The set-up will be at a millimeter scale to be used during surgery and will combine four different modalities, two quantitative (spectral and lifetime measurements) and two qualitative (fluorescence and second harmonic generation imaging). The use of a non-linear excitation, with an accordable femtosecond laser in the near-infrared, solves some of the drawbacks of a confocal set-up with visible excitation. Under this nonlinear excitation there is no overlapping between emission and excitation, a better penetration depth can be achieved and photodamages are more localized and controlled. In addition, working in the tissue therapeutic window, the access of two optical modalities, the fluorescence and the non-linear phenomenon of second harmonic generation (SHG), and the accordable cavity offers the possibility to excite a large range of molecules. This system can also be based on tissues autofluorescence to get information on its microenvironment in real time without any exogenous dye. Endogenous fluorescence was first considered as a noise in the data, as a result of a background signal compared to the emission of exogenous markers. Recent progresses in microscopy, data analysis, and will to work without markers for a faster integration with surgical process lead to reconsider this judgment, in such a way that autofluorescence in appearing now as a powerful and very promising tool to monitor biological conditions and environment in real time [11]. Literature on the subject is growing and put in light a large number of molecules able to give a fluorescence signal as for example Tyrosine, tryptophan, collagen, NADH, FAD, lipopigments, Porphyrins, Keratin, Elastin. In the visible and near infrared excitation range, five major molecules have been highlight in brain tissues: NADH, FAD, lipopigments, porphyrin and chlorin [12] and studied previously on rat slices and human biopsies [13–15].

In parallel to the technical development of such endomicroscope, it's important to well characterize the endogenous fluorescence under a large scale of excitation wavelengths and through the different modalities to be implemented. Consequently, a database of the optical signature for the different brain tissues and tumors has to be implemented in order to be able to establish indicators of tumor cell at a high resolution.

To get vital information but still poorly documented in the literature such as variation of this signal with time, an endoscopic system has been installed at the Saint-Anne Hospital (Paris, France), as close as possible to the surgery room allowing to analyze biopsies just after surgeries. This system is a fibered endoscope working with a visible excitation and measures the spectral emission and fluorescence lifetime of human brain samples. Since usually there is a significant time lapse between tumor resection and optical analysis due to the selection and transport of the sample between the operating room and microscopy assembly. Even during neurooncology interventions, this time delay still exists: the operation lasts often several hours and biopsies can be taken at any moment, but only brought to the anatomopatholgy service at the end of it.

The aim was to record spectral and lifetime measurements from different endogenous molecules from human brain sample and to follow the evolution of the fluorescence response within time after excision. Samples were provided from adult patients operated in the Sainte Anne Hospital Neurosurgery Department (Paris, France), longue been known for its activity's in cancer treatment and for its collaborations with laboratory's and hospitals. Surgery room in Sainte Anne is close to the experimental setup location allowing to ensure a minimum of time lag between the biopsy collection and the sample examination under the optical endoscope. This study was accomplished on metastasis and epileptic cortex tissues.

10.2 Experimental Section

10.2.1 Optical Assembly

A detailed description of the experimental set-up, previously published [16], for consistency it will be transcribed here. The optical experimental setup is schematized in Fig. 10.1. Summarizing, the excitation source is composed by two PicoQuant separated laser diodes coupled with a shutter, which emit 70 ps pulses centered at 405 nm (LDH-P-C-405B) and 375 nm (LDH-P-C-375B) with a 40 MHz repetition rate. The laser beam is coupled into a SEDI ATI Fibres Optiques (HCG M0200T) optical fiber, specifically dedicated to the excitation source. It is a 200 µm of core diameter a 0.22 numerical aperture (NA) multimode optical fiber, preceded by an injector coupled with a band pass filter centered at 375 or 407 nm. The average power at the fiber output is less than 1 mW. The overall spatial resolution was of 500 µm [17]. The fluorescence signal is collected by a multimode fiber (HCG M0365T), 365 µm core diameter and 0.22 NA, via a collimator at its proximal output coupled with a high pass filter. The collected fluorescence is separated in two through a beam splitter towards two detectors. A cooled spectrometer (QP600-1-UV-VIS, Ocean Optics) was used for spectral measurements., The collected fluorescence was guided to A photomultiplier tube (PMT) (PMA-182 NM, PicoQuant GmbH, Berlin, Germany) was used for time resolved measurements. The PMT temporal resolution was of 220 ps. The synchronization output signal from the diode driver and the start signal from the PMT were connected to their respective channels on the data acquisition board Time-Correlated Single Photon Counting (TCSPC) (Time-Harp 200, PicoQuant GmbH, Berlin, Germany). The selection of spectral emission band was carried out through a motorized filter wheel (FW102C, Thorlabs, Newton, USA), placed in front of the PMT. Five filters were used at the 405 nm excitation wavelength (Semrock, New York, USA): 450 ± 10 nm, 520 ± 10 nm, 550 ± 30 nm, 620 ± 10 nm and 680 ± 10 nm corresponding respectively to the five endogenous fluorophores: reduced Nicotinamide adenine dinucleotide (NADH), flavin (FAD), lipopigments (Lip), porphyrin (Porph) and chlorin. At the 375 nm excitation wavelength, only the 450 ± 10 nm and the 520 ± 10 nm filters were used. The time of life and spectroscopic measurements were carried out on the same setup and the time required to measure each fluorophore lifetime was two seconds.

A Matlab software script was developed to process the spectral measurements namely to fit the different endogenous molecules that emit fluorescence in brain tissues. The results of a fitted spectrum are presented in Fig. 10.2, where the emission spectra of control group at 405 nm wavelength are displayed [18, 19]. It basically

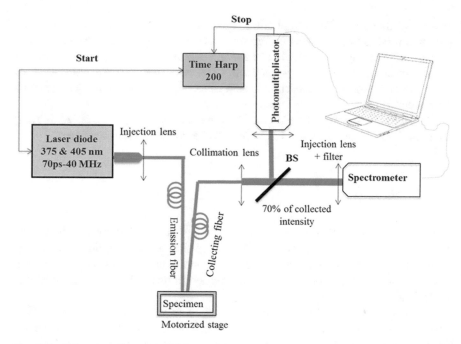

Fig. 10.1 Illustration of the optical setup used for spectral and lifetime fluorescence measurements of human brain tissues

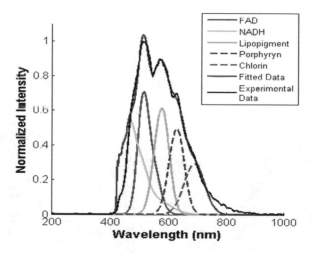

Fig. 10.2 Spectrum of a control tissue with a 405 nm excitation wavelength fitted with a homemade Matlab script [18, 19]

shows that at this wavelength one can excite correctly the five molecules mentioned, and by extracting the fitting curve of each molecule its maximum intensity can be followed overtime easily. The Symphotime (PicoQuant, GmbH, Berlin, Germany) software was used to treat fluorescence lifetime data, collected by the same equipment.

For XY scanning a dedicated mechanical support was mounted on a motorized micro translator stage (Thorlabs, Newton, USA) was used. The scanning velocity in X-dimension was 100 μm/s and the acquisition time during X-line scanning was three seconds per fluorescence spectrum.

The optical set-up was placed in the Neuropathology Department of Sainte Anne Hospital (Paris, France), to be as close as possible to the in vivo conditions.

10.2.2 Samples

The study here under consideration was carried out with the approval of the Institutional Review Board of Saint Anne Hospital (CPP S.C. 3227), the adult patients sign an agreement before surgery and under these conditions the anatomopathologist is allowed to supply a fraction of the resected biopsy for research purpose directly after the operation.

Samples came from adult patients submitted to surgery in the Sainte Anne Hospital Neurosurgery Department (Paris, France). Five samples have been analyzed: three metastasis samples (tumor samples) and two epilepsy surgery samples (control sample). Each sample was selected by a senior pathologist on a fresh resected specimen.

Samples were analyzed if there was enough material for gold-standard histopathology and if the resected specimen was representative of a tumor or a healthy tissue. A resected sample of metastases and a typical histological staining of this group is shown in Fig. 10.3. On this figures three regions can be distinguished, a healthy part of the cortex, where control biopsies were taken, a meninx wraps

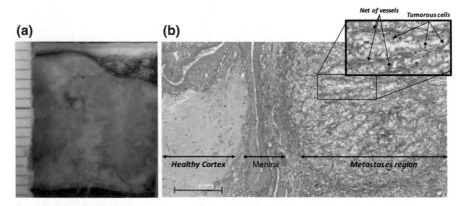

Fig. 10.3 **a** Image of a metastases biopsy on the set-up just before measurements. **b** Histological gold standard Hematoxylin and Eosin (H&E staining) of a metastases biopsy

between region of cortex, and a metastasis area, very identifiable by the very dense
vascularization creating a net around the tumorous cells.

10.3   Results

10.3.1   Spectral Emission

Just for reference, fluorescence dynamics reference times will be after excision (t_0),
one hour after excision (t_1), two hours after excision (t_2) and three hours after excision
(t_3) and so on. Accordingly, a slight decrease within the first 3 h was noticed in
fluorescence intensity and a strong decrease (>50%) of the fluorescence intensity
was observed for the longest time intervals (t_4 and t_5).

For the five investigated endogenous fluorophores, the maximum fluorescence
intensity of the fitted spectrums was lower after three hours (t_3) than at the initial

Fig. 10.4 Normalized
fluorescence intensity
emission of NADH and FAD
for different endoge-nous
molecules of metastasis (**a**)
and Control (**b**) tissue using
excitation at 375 nm

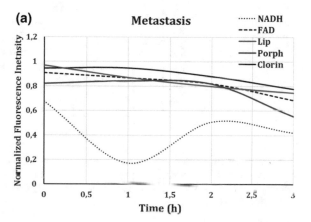

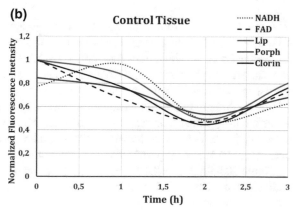

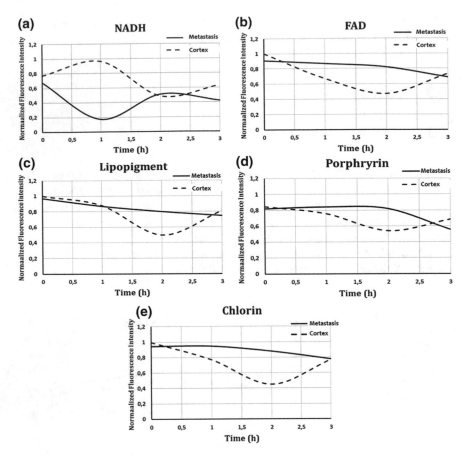

Fig. 10.5 Mean maximum fluorescence intensities at 405 nm excitation of the five endogenous fluorophores: **a** NADH, **b** FAD, **c** lipopigment, **d** porphrin and **e** chlorin for different times. (For reference t_0, after excision, t_1, one hour after excision, t_2 two hours after excision and three t_3, hours after excision)

measurement (t_0). It should be noted that this decrease was not always seen and a large variability has been observed in the dataset. The normalized fluorescence intensities for each endogenous molecule (NADH, FAD, Lipopigments, Porphyrins and Chlorins) for Metastasis and Control tissues are shown in Fig. 10.4. These results were acquired using 375 nm excitation wavelength.

For the 405 nm excitation wavelength, cortex samples revealed stronger fluorescence intensity values than metastasis, behavior which was maintained during all the protocol. At excitation wavelength of 375 nm, the opposite situation seemed to take place (Fig. 10.4) even if partial results did not allow any conclusion.

Figure 10.5 illustrates the endogenous fluorescence variation within time for each endogenous molecule. This study was accomplished for metastasis and healthy cortex tissue. As shown, a different trend was observed within time between tumorous and

Fig. 10.6 The spectral behavior of the emission fluorescence from metastasis sample until the fifth hour and on the same ROI using: 405 nm excitation wavelength top graph and 375 nm excitation wavelength, bottom graph

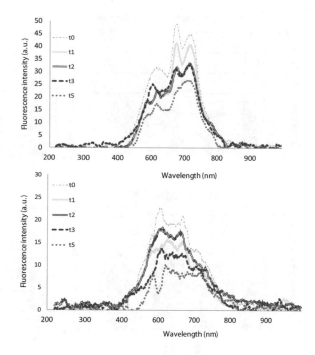

healthy tissue. In both case and for all endogenous molecules, a remarkable change in fluorescence intensity was recorded after 2 h.

To explore changes in the pattern of the spectra, the last metastasis sample on the same ROI, were recorded until the fifth hour. The results, shown in Fig. 10.6, revealed that spectral pattern is also affected under 405 nm excitation wavelengths as follows: at t_0, three emitted peaks were registered around 600, 680 and 700 nm.

At 375 nm excitation wavelength, the two first peaks (600 and 680 nm) were also present. After five hours (t_5) the spectral shapes did not show any of these peaks for both 375 and 405 nm excitation wavelengths. These changes could be associated with tissue oxygenation.

10.3.2   Lifetime Measurements

Figure 10.7 summarizes the lifetime evolution of fluorescence measurements for the three-metastasis samples and the two samples of healthy human cortex using 405 nm excitation wavelength.

The same trend was observed for the three-metastasis samples: lifetime values slightly shortened with time at 405 nm excitation wavelengths. At 375 nm excitation wavelength longer lifetime values have been found, for the 450 ± 10 and 520 ± 10 nm filters with regard to 405 nm excitation wavelengths values. The lifetime values found

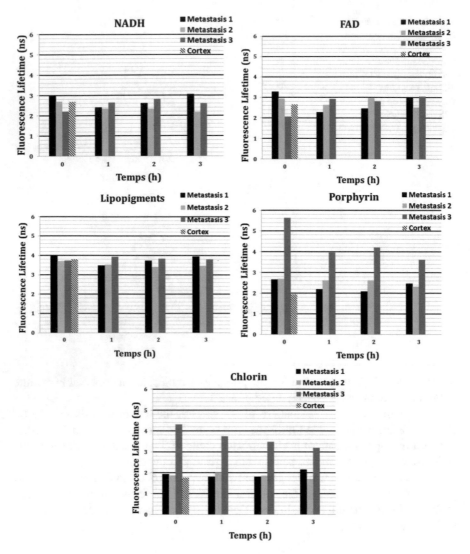

Fig. 10.7 Fluorescence lifetimes dynamics of the five endogenous fluorophores obtained with 405 nm excitation wavelength: after excision (t_0), one hour after excision (t_1), two hours after excision (t_2) and three hours after excision (t_3)

were in the 1.69–5.63 ns range. It should be noted that at 405 nm, the values for the 620 ± 10 and 680 ± 10 nm filters give the feeling to be longer for metastasis samples than for cortex samples. The differences in fluorescence lifetime could be associated with tissue viability. In fact, after three hours the cellular structure did changed as it can be noticed on histological analysis taken right and after the surgery, i.e. after t_3. Therefore the tissue microenvironement also changes, which likely affects lifetime measurements. Generally, three hours after tissue excision, no clear differences have been observed on fluorescence lifetime measurements between all investigated fluorophores on these metastasis samples.

Shorter lifetimes seemed to take place in healthy cortex samples when compared with those of metastasis samples particularly for the last two filters, namely 620 ± 10 and 680 ± 10 nm. This is likely to be associated with porphyrin and chlorin behaviors, at 405 nm excitation wavelength.

10.4 Discussions and Conclusions

The characteristics of variation of endogenous fluorophore in human tissues over time were addressed here for the first time. Five fluorophores were considered in this study: reduced nicotinamide adenine dinucleotide (NADH) and flavins (FAD), two coenzymes in the Kreps cycle. Each plays major roles in energetic metabolisms, such as glycolysis, are the most investigated endogenous fluorophores, often look at to express the redox ratio between them [20–22], lipopigments in which the fluorescence emission depends on the oxidation degree of its components [23], porphyrin which has a relative increase in fluorescence in tumorous tissues [24] and Chlorin. These fluorophores were excited at 375 and 405 nm wavelengths and fluorescence emission was observed between 400 and 900 nm, Excitation at 405 nm wavelength was found to be the most efficient to excite these five molecules, while 375 nm excitation was seen more specific for NADH and FAD. Working with autofluorescence signals, for long considered as noise, offers major advantages as not required potential bias due to the introduction of external molecules the tissue molecules are directly being analysed, and in a long term consideration it allows to avoids drawbacks related with the ministration of external agents to a patient. However, it is not an easy signal to work with, mainly due to its very low intensity, thus requiring precise measurement techniques to extract it correctly and a lot of signal processing to be performed in order to extract meaningful information. If endogenous fluorescence is an argument for cost reduction, the choice of the modality to observe it is also driven by a low cost will. Spectral portal devices have been evaluated at a cost of 15,000 US dollars, a ship technology in medical imaging and its small size is also very suitable to the operating room [25]. Moreover spectroscopy is a very fast measurement technique, with a high resolution and very easy to implement. Preliminary results of combined spectral with quantitative measurements, as fluorescence lifetime, has already been reported [13–15, 18]. Fresh and fixed extracted slices from rat brain previously injected tumorous cell lines [13] and fixed human tissues have been measured to discriminate meningioma grades I and II [14] and, in the latest, fresh human samples minutes after operation started to be analysed [19]. These preliminary results have boosted to further studies on fresh biopsies over time, to see how the samples will change and thus to see if bringing them to a multiphoton platform an hour away from the hospital will still give viable outcomes. Presented results showed that autofluorescence decreased with time after extraction; however this decrease was seen to be high for fluorescence intensity but not strong for lifetime measurements. This behaviour is consistent with previous published results on autofluorescence variation with time [26]. The same trend was observed for Metastasis and control samples (cortex providing from epilepsy surgery): a slight decrease in fluorescence intensity and lifetime values with time. Then, this change is general for tissues and not dependent on its nature. The fluorescence signal starts to have a significant change in shape after two hours and loses most of its distinct emission peak after five hours. Analysis on resected tissues should happen in the next two hours to give a result close to what the surgeon will see with an endoscopic probe during intervention. In anatomopathology samples are stored for several days to assess a diagnosis. To conserve the tissues, a fixation with formaldehyde is performed. An interesting experiment following this study will be to fix a tissue after each significant time ($t_0, t_1, t_2, \ldots$) and measure its spectrum to see the real impact of fixation and if the moment of fixation plays a role in the spectral response. This information will help to validate or not measurements made on a fixed tissue on multiphoton microscope hours away from the surgical room, in order to ingrate them in an optical database. This study has shown that fluorescence lifetime stays quite stable through time even at low fluorophores concentration. It has already been reported in the literature that this technique is not sensitive to molecules concentration, excitation wavelength, setup type, but only to environmental and structural factors. Fluorescence lifetime is then influenced for example by variation of pH, temperature, solvent, media viscosity, molecule interaction, microenvironement of the molecules and molecule structure [27–30]. This gives a perfect pendant to classical techniques to monitor fluorescence such as spectroscopy, which gives quantitative values only on fluorescence intensity, a parameter highly influenced by the concentration and system probe geometry. Fluorescence lifetime brings a real quantitative value insensitive to any experimental factors, however the field is still very new and any conformational change in the molecules could result in a change in the lifetime values, which results in a difficult data interpretation. Nevertheless recent literature report some ground breaking results on differentiating tumors from healthy brain tissue [8, 21, 31]. This work is still preliminary with only two types of tissues and few samples within the types. In order to have a more statistical result concerning the evolution of fluorescence signal with time on resected samples, a larger cohort is necessary to be examined. This cohort should include different new tumorous groups, at least ten of each, such as glioma, a very deadly tumor in its high grade, and meningioma the most common brain tumor in adult population [5, 16] and also infiltrated boundaries of the different tumour types. Of course, these studies on tumorous tissues should be compared to the corresponding healthy ones.

Although the focus addressed here has been the evolution of signal over time, it also gives the possibility to look at the difference between a metastasis signals and a healthy control signal. In fact, an increase in fluorescence intensity at 405 nm excitation wavelength was highlight from metastasis samples to their control during all the protocol, on four of five molecules. This result shows the promising future of spectral analysis in brain tumor diagnosis, giving a first insight on how to discriminate a tumorous tissue. However, this indicator doesn't seem so robust concerning to key parameters such has excitation wavelength, or molecules environment. Spectral analysis as the only mean of contrast appears limited to tackle the difficult problem of distinguishing healthy boundaries from infiltrated tumor cells. The use of a multimodal approach to overcome this limit by combining spectral analysis with other modalities, either more quantitative such as lifetime measurements, or more qualitative such SHG images will be for sure an added value [31].

Acknowledgements This Work as a part of the MEVO project was supported by "Plan Cancer" program founded by INSERM (France), by CNRS with "Défi instrumental" grant, and the Institut National de Physique Nucléaire et de Physique des Particules (IN2P3). Thanks to the PIMPA Platform partly funded by the French program "Investissement d'Avenir" run by the "Agence Nationale pour la Recherche" (grant "Infrastructure d'avenir en Biologie Santé – ANR – 11-INBS-0006").

References

1. J.M. López Piñero, Nine centuries of cranial surgery. Lancet **354**(Suppl), SIV35 (1999)
2. K. Uluç, G.C. Kujoth, M.K. Başkaya, Operating microscopes: past, present, and future. Neurosurg. Focus **27**, E4 (2009). https://doi.org/10.3171/2009.6.FOCUS09120
3. K.W. Li, C. Nelson, I. Suk, G.I. Jallo, Neuroendoscopy: past, present, and future. Neurosurg. Focus **19**, E1 (2005). https://doi.org/10.3171/foc.2005.19.6.2
4. G. Unsgaard, O.M. Rygh, T. Selbekk, T.B. Müller, F. Kolstad, F. Lindseth, T.A.N. Hernes, Intraoperative 3D ultrasound in neurosurgery. Acta Neurochir. (Wien) **148**, 235–253; Discussion 253 (2006). https://doi.org/10.1007/s00701-005-0688-y
5. Q.T. Ostrom, H. Gittleman, J. Xu, C. Kromer, Y. Wolinsky, C. Kruchko, J.S. Barnholtz-Sloan, CBTRUS statistical report: primary brain and other central nervous system tumors diagnosed in the United States in 2009–2013. Neuro-Oncol. **18**, v1–v75 (2016)
6. D.A. Hardesty, N. Sanai, The value of glioma extent of resection in the modern neurosurgical era. Front. Neurol. **3**. https://doi.org/10.3389/fneur.2012.00140
7. N. Sanai, M.S. Berger, Glioma extent of resection and its impact on patient outcome. Neurosurgery **62**, 753–766 (2008) https://doi.org/10.1227/01.neu.0000318159.21731.cf
8. P.V. Butte, A.N. Mamelak, M. Nuno, S.I. Bannykh, K.L. Black, L. Marcu, Fluorescence lifetime spectroscopy for guided therapy of brain tumors. NeuroImage **54**, S125–S135 (2011). https://doi.org/10.1016/j.neuroimage.2010.11.001
9. A.H. Zehri, W. Ramey, J.F. Georges, M.A. Mooney, N.L. Martirosyan, M.C. Preul, P. Nakaji, Neurosurgical confocal endomicroscopy: a review of contrast agents, confocal systems, and future imaging modalities. Surg. Neurol. Int. **5**, 60 (2014). https://doi.org/10.4103/2152-7806.131638
10. P.A. Valdés, D.W. Roberts, F.-K. Lu, A. Golby, Optical technologies for intraoperative neurosurgical guidance. Neurosurg. Focus **40**, E8 (2016). https://doi.org/10.3171/2015.12.FOCUS15550
11. A. Chorvatova, D. Chorvat, Tissue fluorophores and their spectroscopic characteristics, ed. by L. Marcu, P. French, D. Elson, *Fluorescence Lifetime Spectroscopy and Imaging* (CRC Press, 2014), pp. 47–84
12. A.C. Croce, G. Bottiroli, Autofluorescence spectroscopy and imaging: a tool for biomedical research and diagnosis. Eur. J. Histochem. **58** (2014). https://doi.org/10.4081/ejh.2014.2461
13. D.A. Haidar, B. Leh, M. Zanello, R. Siebert, Spectral and lifetime domain measurements of rat brain tumors. Biomed. Opt. Express **6**, 1219–1233 (2015). https://doi.org/10.1364/BOE.6.001219

14. M. Zanello, F. Poulon, P. Varlet, F. Chretien, F. Andreiuolo, M. Pages, A. Ibrahim, J. Pallud, E. Dezamis, G. Abi-Lahoud, F. Nataf, B. Turak, B. Devaux, D. Abi-Haidar, Multimodal optical analysis of meningioma and comparison with histopathology. J. Biophotonics (2016). https:// doi.org/10.1002/jbio.201500251

15. M. Zanello, A. Ibrahim, F. Poulon, P. Varlet, B. Devau, D. Abi Haidar, in *Proceedings of the 4th International Conference on Photonics, Optics and Laser Technology, PHOTOPTICS*, vol. 1 (2016), pp. 13–17

16. D. Abi Haidar, B. Leh, A. Allaoua, A. Genoux, R. Siebert, M. Steffenhagen, D.A. Peyrot, N. Sandeau, C. Vever-Bizet, G. Bourg Heckly, I. Chebbi, Spectral and lifetime domain measurements of rat brain tumours, in *SPIE (Ed.), Spectral and Lifetime Domain Measurements of Rat Brain Tumours* (San Jose, United States, 2012), p. 82074P. https://doi.org/10.1117/12.907411

17. B. Leh, R. Siebert, H. Hamzeh, L. Menard, M.-A. Duval, Y. Charon, D. Abi Haidar, Optical phantoms with variable properties and geometries for diffuse and fluorescence optical spectroscopy. J. Biomed. Opt. **17**, 108001 (2012). https://doi.org/10.1117/1.JBO.17.10.108001

18. A. Ibrahim, F. Poulon, R. Habert, C. Lefort, A. Kudlinski, D.A. Haidar, Characterization of fiber ultrashort pulse delivery for nonlinear endomicroscopy. Opt. Express **24**, 12515–12523 (2016)

19. A. Ibrahim, F. Poulon, F. Melouki, M. Zanello, P. Varlet, R. Habert, B. Devaux, A. Kudlinski, D. Abi Haidar, Spectral and fluorescence lifetime endoscopic system using a double-clad photonic crystal fiber. Opt. Lett. **41**, 5214 (2016). https://doi.org/10.1364/OL.41.005214

20. G. Papayan, N. Petrishchev, M. Galagudza, Autofluorescence spectroscopy for NADH and flavoproteins redox state monitoring in the isolated rat heart subjected to ischemia-reperfusion. Photodiagnosis Photodyn. Ther. **11**, 400–408 (2014). https://doi.org/10.1016/j.pdpdt.2014.05.003

21. V.K. Ramanujan, J.-H. Zhang, E. Biener, B. Herman, Multiphoton fluorescence lifetime contrast in deep tissue imaging: prospects in redox imaging and disease diagnosis. J. Biomed. Opt. **10**, 051407 (2005). https://doi.org/10.1117/1.2098753

22. M.C. Skala, K.M. Riching, A. Gendron-Fitzpatrick, J. Eickhoff, K.W. Eliceiri, J.G. White, N. Ramanujam, In vivo multiphoton microscopy of NADH and FAD redox states, fluorescence lifetimes, and cellular morphology in precancerous epithelia. Proc. Natl. Acad. Sci. **104**, 19494–19499 (2007)

23. S. Takehana, M. Kaneko, H. Mizuno, Endoscopic diagnostic system using autofluorescence. Diagn. Ther. Endosc. **5**, 59–63 (1999)

24. M. Wolman, Lipid pigments (chromolipids): their origin, nature, and significance. Pathobiol. Annu. **10**, 253 (1980)

25. S.A. Toms, W.-C. Lin, R.J. Weil, M.D. Johnson, E.D. Jansen, A. Mahadevan-Jansen, Intraoperative optical spectroscopy identifies infiltrating glioma margins with high sensitivity. Neurosurgery **61**, 327–336 (2007). https://doi.org/10.1227/01.neu.0000279226.68751.21

26. A.C. Croce, S. Fiorani, D. Locatelli, R. Nano, M. Ceroni, F. Tancioni, E. Giombelli, E. Benericetti, G. Bottiroli, Diagnostic potential of autofluorescence for an assisted intraoperative delineation of glioblastoma resection margins. Photochem. Photobiol. **77**, 309–318 (2003)

27. M.Y. Berezin, S. Achilefu, Fluorescence lifetime measurements and biological imaging. Chem. Rev. **110**, 2641–2684 (2010). https://doi.org/10.1021/cr900343z

28. A.S. Kristoffersen, S.R. Erga, B. Hamre, Ø. Frette, Testing fluorescence lifetime standards using two-photon excitation and time-domain instrumentation: rhodamine b, coumarin 6 and lucifer yellow. J. Fluoresc. **24**, 1015–1024 (2014). https://doi.org/10.1007/s10895-014-1368-1

29. L. Marcu, Fluorescence lifetime techniques in medical applications. Ann. Biomed. Eng. **40**, 304–331 (2012). https://doi.org/10.1007/s10439-011-0495-y

30. A.G. Ryder, S. Power, T.J. Glynn, J.J. Morrison, Time-domain measurement of fluorescence lifetime variation with pH, in *BiOS 2001 The International Symposium on Biomedical Optics. International Society for Optics and Photonics* (2001), pp. 102–109

31. P.P. Provenzano, K.W. Eliceiri, P.J. Keely, Multiphoton microscopy and fluorescence lifetime imaging microscopy (FLIM) to monitor metastasis and the tumor microenvironment. Clin. Exp. Metastasis **26**, 357–370 (2009). https://doi.org/10.1007/s10585-008-9204-0

Chapter 11
Photochromic Materials Towards Energy Harvesting

Gonçalo Magalhães-Mota, Pedro Farinha, Susana Sério, Paulo A. Ribeiro and Maria Raposo

Abstract Solar to electrical energy conversion is one of the possible approaches regarding energy demand and sustainability. While traditional semiconductor photovoltaic units are seen to be well implemented in the market, with companies offering packs for both domestic and industries, the search for new materials and technologies aiming improvement in efficiency together with low cost issues has been continuously stressed over the past 20 years. This work point out a novel approach for energy conversion and storage devices, based on the so called photochromic electrets or photoelectrets, by using materials containing highly polarisable molecules as multifunctional diarylethenes (DTEs) and azobenzenes. This is supported by experimental results which show that (1) DTE-CN blends after electrical field induced orientation are able to induce external current response when submitted to visible light pulse (2) visible light is able to induce birefringence in layer-by-layer (LbL) films of poly{1-(4-(3-carboxy-4-hydroxy-phcnylazo) benzenesulfonamido)-1,2-ethanediyl, sodium salt}(PAZO) and poly (allylamine hydrochloride) (PAH). These preliminary results allow to conclude that solar devices based on photoelectrets are worth to be further investigated towards energy harvesting.

G. Magalhães-Mota · P. Farinha · S. Sério · P. A. Ribeiro · M. Raposo (✉)
CEFITEC, Departamento de Física, Faculdade de Ciências e Tecnologia,
Universidade Nova de Lisboa, 2829-516 Caparica, Portugal
e-mail: mfr@fct.unl.pt

G. Magalhães-Mota
e-mail: g.barreto@campus.fct.unl.pt

P. Farinha
e-mail: p.farinha@campus.fct.unl.pt

S. Sério
e-mail: susana.serio@fct.unl.pt

P. A. Ribeiro
e-mail: pfr@fct.unl.pt

© Springer Nature Switzerland AG 2018
P. A. Ribeiro and M. Raposo (eds.), *Optics, Photonics and Laser Technology*, Springer Series in Optical Sciences 218, https://doi.org/10.1007/978-3-319-98548-0_11

11.1   Introduction

The depletion of fossil fuels reserves and greenhouse gases suppression are two key issues to be solved concerning energy demand and sustainability [1]. Therefore, it extremely urgent to spread and encourage the use of renewable energy sources, such as sun, wind and water, especially in what concerns to electrical energy production. Since normally renewable energies sources are not accessible when they are more needed it is also fundamental to develop efficient energy storage systems at low cost. The conversion of solar energy into electricity is one of the possible alternatives however it is still somehow costly for large scale applications.

Sunlight hits the earth's surface intermittently and with unpredictable intensity, consequently the production of electric energy from solar radiation can change rapidly, making it difficult to distribute and thus requiring storage units. However expectations can change regarding micro-power generation applications.

Advances in nanotechnology and generally in materials science may prompt new solutions for efficient and low cost solar energy conversion-storage systems [2]. For example, the new electrochemical capacitors, called supercapacitors [3], are a promising alternative to the most commonly used lithium ions, which lose their charging capacity with successive charge-discharge cycles, and therefore, not meeting the high speed requirements for high power systems. These systems exhibit important advantages such as high energy densities stored, fast charge-discharge cycles and higher service life and thus are expected to become the next power generation storage devices [4, 5].

Another approach to develop energy storage devices is the use of materials containing highly polarisable molecules, which have been adequately oriented in the medium to attain a net polarization. Electrical energy can then be delivered to an external circuit by making polarization to change over its steady value, which can be implemented either through piezoelectric or pyroelectric effects our even, making use of light sensitive molecular dipoles, which undergo chemical reaction triggered by light. This concept of the electret is not new and is well accounted in a book by Sessler [6], where an extensive review and some examples are given. Herein will be reviewed the techniques, concepts and materials regarding electrets towards the development of energy harvesting devices focused on light to electrical energy transduction.

11.2   Energy Harvesting Electrets

Electret is an old concept, consisting of a dielectric material having a permanent electric charge or polarization, capable to generate an external electric field. The term electret was due to Oliver Heaviside in 1885 [7] in analogy to the magnet. Research in dielectric electrets became popular with the development of the electret microphone in 1962 by West and Sessler [8]. Charge can be accumulated in certain

(a) **(b)**

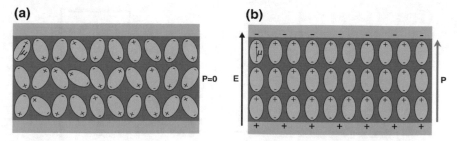

Fig. 11.1 Schematization of the poling of a dipolar electret for achieving a net polarization: **a** unpolarised electrets with random aligned dipoles and zero net polarization; **b** polarised medium with all dipoles oriented in the same direction giving rise to a net polarisation P

dielectric materials to become an electret in charge traps either in surface or bulk, or through bulk oriented electric dipoles, Fig. 11.1.

Once polarised, the electrical energy is stored in the material and can be converted into electrical energy, for example, to an external circuit if one acts in the medium so that a change in polarization is achieved.

Examples of most investigated electrets are the carnauba wax, in early stages, [9], polyethylene (PE), polypropylene (PP), fluoroethylenepropylene (Teflon-FEP) [10, 11] and poly (vinylidene fluoride) (PVDF) [12–15].

Particular interest has been paid to ferroelectric electrets, which also present piezo-electric and pyroelectric properties that can be directly used for energy harvesting, just by inducing polarization changes through stress/strain application or temperature changes. This can be clearly seen in following Mopsik model for the electrical polarization arising from a medium with N non interacting molecular dipoles of dipole moment μ, placed at the centre of spherical cavities of known dimension, wrapped in a dielectric medium [16], as sketched in Fig. 11.2. Under these conditions the electrical net permanent polarization of a planar dielectric medium can be written in terms of molecular dipole moment and average dipole orientation with respect to dielectric surface and can be determined by the following formula:

$$P = \frac{N}{V} \frac{\varepsilon_\infty - 1}{3} \mu \langle \cos \theta \rangle \tag{11.1}$$

where V is the volume of the spherical cavities, ε_∞ is the dielectric constant at elevated frequencies, where dipoles no longer follow the electric field and $\langle \cos \theta \rangle$ the average dipole moment orientation.

Accordingly, a change in medium polarization can arise from change in dipole concentration, dipole moment or dipole orientational order, as changes in dielectric constant are considered negligible. These changes, for example with respect to external stimulus "X", can be expressed by:

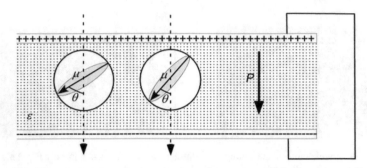

Fig. 11.2 Schematization of molecular dipoles placed inside of spherical cavities wrapped in a dielectric medium according with Mopsik model for the calculation of net permanent polarization P

$$\frac{\partial P}{\partial X} = \frac{N}{V^2} \frac{\varepsilon_\infty + 2}{3} \mu \langle \cos \theta \rangle \frac{\partial V}{\partial X} + \frac{\varepsilon_\infty + 2}{3} \langle \cos \theta \rangle \frac{\partial \mu}{\partial X} + \frac{N}{V} \frac{\varepsilon_\infty + 2}{3} \mu \frac{\partial \langle \cos \theta \rangle}{\partial X}$$
(11.2)

Moreover, changes in polarization gives rise to surface charge imbalance and thus to an electrical current if connected to an external circuit. For example, "X" could be stress or strain, temperature or electric field including that of light. This effect has been reported for energy harvesting in fluoroethylenepropylene ferroelectrets and cross-linked polypropylene through piezoelectrics effect [17, 18]; stacked piezoelectrets clamped to a seismic mass have been proposed by Podrom et al. [19], sandwiched fluoroethylene propylene films with charged parallel voids between, have been proposed by Zhang et al. [20]. Also Cuadras suggested pyroelectric cells based on screen printed lead zirconate titanate (PZT) and PVDF to provide energy to low power autonomous sensors [21].

11.3   The Photochromic Electret

As mentioned above the change in polarization in a dielectric medium can be triggered by light, for example, through the change in molecular dipole moments as a result of some photochemical reaction, which alters the molecule or its conformation. Generally speaking, the layout of a molecular dipole adequate to interact with light, consists of a donor acceptor groups bonded by a π conjugated bond, as sketched in Fig. 11.3.

An example of these, are the so called multifunctional diarylethenes (DTEs) derivatives, known since the eighties as molecular switches. These photo responsive molecules present not only two isomer forms, exhibiting different conformation, but also different dipole moments. In addition, switching from conformation can be induced by light, with visible light promoting the opening of aromatic ring while UV-light promotes its closure, Fig. 11.4.

Under this line Castagna et al. [22] proposed the use of 1,2-bis-[2-methyl-5-(p-cyanophenyl)-3-thienyl]perfluorocyclopentene (DTE-CN), Fig. 11.5, for the creation of photochromic electrets to be used in light energy harvesting devices.

The device will directly convert the photon energy into electric energy thanks to a polarization modulation due to the change in dipole moment caused by the photoreaction. The DTE-CN with its dipole moments towards the fluorine atoms, are colourless in the open ring form and coloured in the closed ring form. Dipole moments differ from 0.81 D in the open ring form to 0.18 D in closed ring form.

For the device a planar capacitor filled with a dielectric material doped with DTE-CN was proposed, thus featuring different colours and dipole moments according with wavelength impinging the device.

Recently, Castagna et al. [22] showed that electrets based in photochromic molecules can be used for energy harvesting, as well as proposed the mechanism for light energy conversion. The device consisted of a parallel plate capacitor filled with a dielectric material doped with a photochromic compound, which features different dipole moments in the colourless and coloured form. The alignment of the dipolar chromophores by poling, so that the photochromic molecules become arranged with the dipole moment perpendicular to the capacitor plates, allowed achieving permanent polarization and thus external electrical field. By irradiating this capacitor device with pulses of both ultraviolet (375 nm) and visible (532 nm) light, an external current was successfully obtained in accordance to forward and backward photochemical reactions. Moreover, current densities of 0.7 pA/mm^2 have been obtained with DTE-CV capacitor photoelectrects, when irradiated by UV light at 375 nm.

Other interesting molecular systems that can be used for light to electrical energy conversion are the azobenzene derivatives. According with general layout for improved light interaction, sketched in Fig. 11.3, the best design for azobenzene derivatives will be the one represented in Fig. 11.6.

Here the conjugation is kept through the benzene rings and the dipolar character of the molecule is conferred by the donor acceptor groups placed at opposite benzene rings. Typical dipole moments can reach 9–10 D and one can even get additional optical response from the photo-isomerization capabilities through the –N=N– azo group, as schematized in Fig. 11.7.

Molecule Charge Transfer Axis

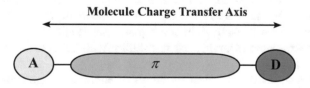

Fig. 11.3 Layout of a molecular dipole suitable for interaction with light electric field consisting of electron donor-acceptor groups bonded through a π bond

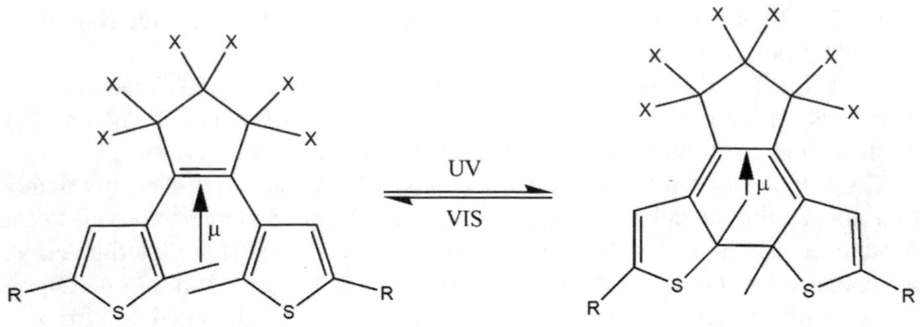

Fig. 11.4 DTE molecule with donor-acceptor group substituent's on opposite sides forming a large dipole, in which the aromatic ring can be opened or closed according with the wavelength of light impinging it

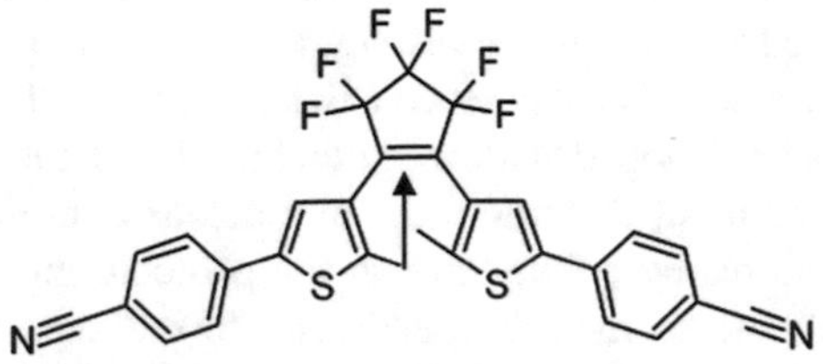

Fig. 11.5 Chemical structures of photochromic molecule 1,2-bis-[2-methyl-5-(p-cyanophenyl)-3-thienyl]perfluoro cyclopentene (DTE-CN). The arrow indicates de direction of molecular dipole moment

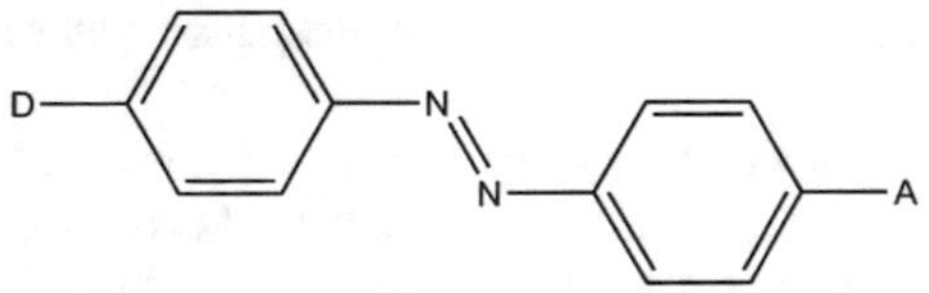

Fig. 11.6 Azobenzene molecule with donor-acceptor group substituents on opposite sides forming a large dipole moment molecule

In fact, azobenzenes have proven to be candidates in the fields of optoelectronics and photonics [23–26]. Azobenzene molecules are characterized by two main absorption bands: a low energy absorption band, in the visible region, and another, at higher energy levels, within the ultraviolet region. In 1937, studies performed by Hartley revealed the photo-isomerization capabilities of these molecules, or the ability to spatially rearrange by light of adequate wavelength. This process can lead to a large macroscopic birefringence in a bulky material containing azobenzene molecules [27].

Figure 11.8 shows the chemical structures of common commercial azobenzene chromophores.

Table 11.1 displays the transition dipole moments and melting point of azobenzene molecules of Fig. 11.8.

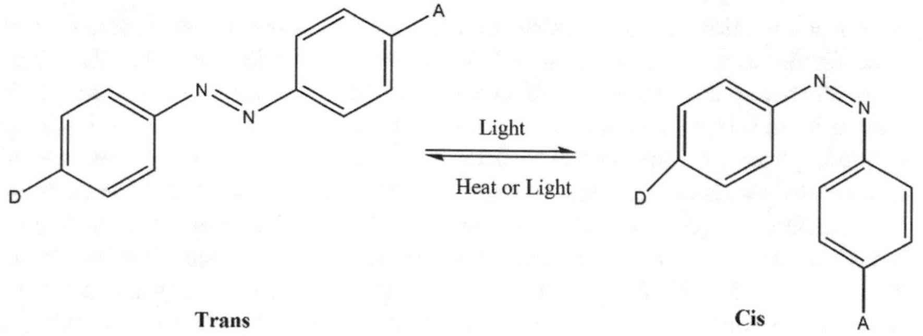

Fig. 11.7 Schematization of photo-isomerization process in azobenzene molecules which can de reached either by adequate wavelength light or heat

Fig. 11.8 Commercial azobenzene chromophores molecules: **a** N-ethyl-N-(2-hydroxyethyl)-4-(4-nitrophenylazo) aniline (Disperse Red 1 – DR1), **b** 2-[4-(2-chloro-4-nitrophenylazo)-N-ethylphenylamino]ethanol (Disperse Red 13 – DR13) and **c** 1-(4-(3-carboxy-4-hydroxy phenylazo) benzene sulfonamido), sodium salt (Disperse Red 19 – DR19)

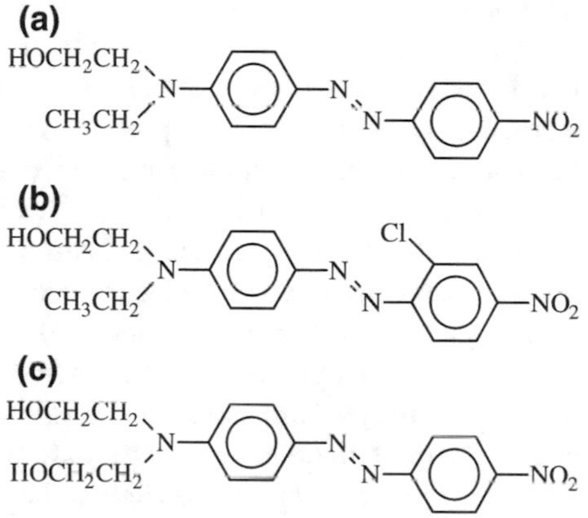

Table 11.1 Physical properties of commercial azo-chromophores: wavelength of maximum absorption in UV-VIS region; transition dipole moment and melting point

Azobenzene chromophore	Maximum absorption wavelength (nm)	Transition dipole moment (D)	Melting point (°C)
DR1	502	8.7	~160
DR13	503	8.6	~130
DR19	495	10	~300

There are some unique features that are worth to mention at this point concerning the photo-isomerisation of azobenzene molecules and consequently materials which incorporate them. The photo-isomerization processes of the azobenzene molecule or group leads to change of the spatial geometric arrangement, through the conversion of isomer form trans to cis (*trans → cis*) as a result of light absorption, or trans to cis (*cis → trans*) induced by light or heat. This photo-isomerization is attained by impinging light of adequate wavelength on the medium, which induces a modification in molecular conformation and molecular reorientation after several *trans−cis−trans* photo-isomerization cycles. As a result of this, the azochromophores will tend to align perpendicular to the polarization direction of the electric field of the incoming light. The geometric spatial change of azobenzenes, and thus, the orientation of the dipole induced by the trans-cis-trans photo-isomerization cycles, allows the reorientation of chromophores, process that even occur at room temperatures, without need of temperature increase in order to improve chromophores mobility within the medium.

The azobenzene chromophores can then be oriented just by using polarized light, through a simple statistical selection process. In fact, as azobenzenes preferably absorb polarized light along their transition dipole axis (the longest axis of the molecule), the probability of absorbing the light varies with $\cos^2\theta$, where θ is the angle between the polarized light and the molecules dipole axis. Hence, the molecules oriented parallel to the polarized light will absorb it, while the molecules oriented perpendicularly will not [28], as shown in Fig. 11.9. In other words for a certain angular distribution of chromophores, some will absorb the polarized light, convert to the cis form and then back to the trans form with a random orientation. The chromophores that become oriented perpendicular to the polarized light will no longer suffer the isomerisation or reorientation process. Thus while the light source remains on, the concentration of chromophores oriented perpendicular to the polarized light electric field, increases constantly until it reaches its maximum saturation level. After this process, a strong birefringence (anisotropy of the refractive index) in the direction perpendicular to the applied electric field and also a dichroism, (anisotropy of the absorption spectrum) have been induced in the medium by light [30].

In addition this photoinduced orientation by polarized light is reversible, i.e., the direction of the chromophores can be altered, by using light with a new polarization angle. The random orientation of the chromophores can be restored by impinging the medium with circular polarized light or heating it, thus restoring the system entropy, even with the non-polarized sunlight being able to steer the chromophores along its irradiation axis [31]. In addition, those molecules are also known to be polarized by sunlight, as is the case of molecules containing azobenzene groups [32].

In looking for a figure of merit it is worth to follow Castagna et al. [22] calculations, which can predict the efficacy of photoelectret devices. By supposing that all dipoles switch instantaneously as a result of the photo-isomerization reaction, the change in polarization ΔP will give rise to a change in external electric field ΔE, the electrical stored energy density in place, U, will be given by

$$U = \frac{\epsilon_0 \Delta E^2}{2} = \frac{1}{2\epsilon_0}(\Delta P)^2 \qquad (11.3)$$

where, ϵ_0 is the electrical permittivity constant $\left(\epsilon_0 = 8.85 \times 10^{-12}\mathrm{C}^2/\mathrm{Nm}^2\right)$. Taking into account expression 1 obtained from Mopsik modelling one can express the energy associated with polarization change by

$$U = \frac{N}{2\epsilon_0 V}\left(\frac{\epsilon_\infty + 2}{3}\langle\cos\theta\rangle\right)^2(\Delta\mu)^2 \qquad (11.4)$$

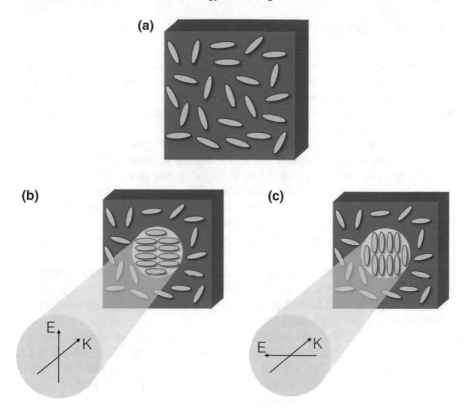

Fig. 11.9 Illustration of the random orientation of the chromophores in the presence of non-polarized light (**a**), and the preferred orientation in the presence of polarized light (**b**, **c**). *E* and *K* correspond to the electric field and wave vectors, respectively. Adapted from [29]

Considering N/V photons with energy $h\upsilon$ equal to that of the photochromic molecules in the electret and ϕ the quantum yield of the photochromic process, one can put the efficiency ratio U/u as

$$\frac{U}{u} = \frac{N\phi}{2\epsilon_0 V h\upsilon}\left(\frac{\epsilon_\infty + 2}{3}\langle\cos\theta\rangle\right)^2 (\Delta\mu)^2 \qquad (11.5)$$

According with Castagna et al. supposing a dipole concentration of 10^{21} cm^{-3}, $\Delta\mu \sim 10$ D, $\langle\cos\theta\rangle \sim 0.9$, $\varepsilon_\infty \sim 2.6$, illuminating the sample with light at 500 nm (2.56 eV), and considering a quantum yield of the photochromic conversion of 0.5, a value of 16% is attained for the "internal" energy conversion efficiency [22].

11.4   Photoelectret Production Details

For making devices it is necessary to prepare films or coatings of these photochromic electrets, which requires two basic steps, namely the preparation of uniform films with the adequate photoabsorption features and to perform the film polarization to achieve a net polarization in the medium. Several techniques can be pointed out for organic thin film production.

The simplest method for prepare these films from organic materials is spill on a solid support a drop of an aqueous or an organic solution of the photochromic material. The solvent is then removed by evaporation, forced or not, and the material stay on the surface without controlled thickness or roughness. The second easier method to prepare thin films with some control of thickness is through the spin-coating method. This method consist of spilling a drop of the solution onto a solid support, which is rotating at a determined angular velocity. The obtained films present regular thickness.

However, there are other techniques of production of organic thin films onto solid supports, which are the Langmuir-Blodgett, self-assembly and layer-by-layer, which allow the preparation of heterostructures of molecular layers with accurate control of thickness and low roughness [33].

The Langmuir-Blodgett technique requires sophisticated equipment, namely a Langmuir trough and respective control units, and amphiphilic molecules, i.e., possessing both a hydrophilic head and a hydrophobic chain. The general preparation procedure of these films is to solubilize the amphiphilic molecules in a volatile solvent, as chloroform, and drop the solution onto the aqueous surface. After the solvent evaporation, the molecules spread over the surface are compressed by a barrier to form the usually called Langmuir monolayer. This monolayer can now be transferred to a solid support by immersing or pulling it in/out the trough. Although this method allows production of well controlled layers, the preparation of these films is expensive, time consuming and not at all suitable for large-scale production.

The self-assembly technique also allows the preparation of molecule layers by chemical adsorption on a solid support and onto other already adsorbed molecule layers. This method requires the use of molecules which have affinity to chemically interact with the solid support and with the next layer of molecules. This limitation severely compromises the use of the technique for large scale production.

The layer-by-layer (LbL) technique consists of alternated physical adsorption of electrically charged polyelectrolytes from aqueous solutions onto solid supports, as schematized in Fig. 11.10. In this technique the solid support is immersed in a policationic solution for a period of time, followed of a quick rinse in water to remove the non-adsorbed molecules; the support, now with a polycationic layer, is immersed in a polyanionic solution for a determined period of time and finally washed again.

A polyelectrolyte bilayer is obtained with this procedure, which can be repeated the number of times equal to the number of desired bilayers. This method of preparation of thin films is cheaper and can be easily adapted for large-scale production. The LbL technique has been used to prepare thin films of azobenzene materials and

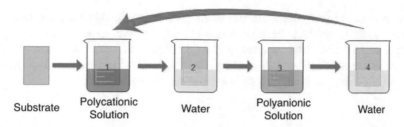

Fig. 11.10 Experimental setup of layer-by-layer technique used to produce nanostructured thin films

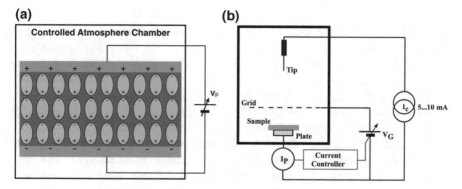

Fig. 11.11 Basic layouts of most common electret poling experimental setups: **a** parallel plate **b** corona triode with current control

its properties have been widely investigated, demonstrating that it can be used for many applications, namely for the application in featured materials [34–38].

The achievement of a net polarization in the photoelectret devices is carried out through a poling process, which can be easily achieved via electrical poling either by parallel plate or corona discharge poling [39], electron beam [40], optical or optical assisted poling [41] or even X-rays and even glow discharge. The most used techniques for electret poling are the parallel plate and corona poling, which are sketched in Fig. 11.11.

Generally, dielectric poling is carried out at elevated temperatures, near to the glass transition temperatures in the particular case of polymeric materials, with the electric field applied to the sample and cooled up to room temperature with applied field. This thermal poling is attained either with an electrode sample in the case of parallel plate configuration, in which an electric potential, V_p, is directly applied, or is carried out through charge injection in the surface of the dielectric material, making the sample electric field to increase and orient the molecular dipoles, Fig. 11.11. These poling techniques can be carried out in normal atmosphere although it might be desirable to perform it in controlled atmosphere. Corona discharge has been used since early stages of electrets, particularly the corona triode configuration. Corona charging does not require the use of electrodes in sample surfaces, which allows to achieve much higher poling electric fields particularly in thin samples. In corona triode a metallic grid has been introduced to improve charge uniformity along the sample surface and better control of poling voltage, V_G. Constant current poling regime by controlling the poling voltage has also been widely used, as it allows full control of poling rates and also makes easier the modelling.

11.5 Birefringence Induction by Visible Light in Azobenzene Containing Materials

As mentioned before the characterization of the effect of light in the creation of anisotropy in azobenzene containing materials is fundamental for the knowledge of electret stabilities and capabilities of creation of large anisotropies and thus causing large changes in electrical polarization. If face of the large changes in orientation that can take place in azobenzene chromophores, as a result of successive *trans-cis-trans* photoisomerisation cycles, the best way to characterize the dynamics of molecular dipole orientation is through optical birefringence measurements.

In Fig. 11.12 is sketched the basic experimental setup that is normally used to study the birefringence dynamics in azobenzene containing materials [42]. The setup consists of a polarized writing beam, normally the Ar^+ laser lines are adequate to induce birefringence in most azochromophores at room temperature, and a probe beam, generally a few mW HeNe laser, to probe birefringence. The writing beam polarization direction can be changed by rotating the half-wave plate with respect to incident beam and when circularly polarized light is required, a quarter-wave plate is place oriented 45° with respect to incident beam. The HeNe laser, 45° polarized with respect to the writing beam, in order to maximize sensitivity, impinges the sample at the same point of writing beam and passes through the analyser, which is initially set to extinguish light before striking the photodetector.

The fraction of light passing through a sample placed between crossed polarisers and reaching the photodetector, will depend then on, the birefringence that is being created by the writing beam, and can be written as

$$\frac{I_v}{I_{v0}} = \left(\frac{\pi \Delta n d}{\lambda} \right)^2 \tag{11.6}$$

where, I_v and I_{v0} are the transmitted and incident light, respectively, Δn is the induced birefringence, d is the thickness of the film and λ is the wavelength. The ratio I_v/I_{v0} is often called the birefringence signal, from which birefringence can be calculated by applying (11.6). A typical creation-relaxation-deletion birefringence curve is presented in Fig. 11.13.

At the start, the probe beam polarizer and analyzer are set to be crossed and the signal is close to zero. At point A, the writing beam is turned on and light starts to pass through the analyser and reach the photodetector as a result of chromophore orientation. The signal then increases to a steady value, which means that no more improvement in dipolar orientation can be achieved. At point B, the writing beam is turned off and birefringence is seen to decay to a residual value, which also corresponds to a residual permanent polarization attained as a result of the disorientation of some of the dipoles. At point C, the writing beam is turned on but with circularly polarized light, causing the signal to go to the starting values at A, which corresponds to the deletion of created birefringence and molecular dipoles orientation will be random.

At this point it is worth to check if visible light following the solar spectrum can induce birefringence in azobenzene molecules. Thin layer-by-layer films of PAH and PAZO polyelectrolytes, Fig. 11.14, supplied by Sigma Aldrich, were prepared from aqueous solutions with a concentration of 10^{-2} M onto fluorine doped tin oxide (FTO) coated glass supports (TEC15, 12–14 Ω/5). The solutions were prepared by dissolving the PAH and PAZO polyelectrolytes in Milli-Q ultrapure water, with resistivity of 18 MΩ cm at 25 °C, (Millipore). Both pH were kept constant to 5 and 7 for PAH and PAZO, respectively. The preparation of the LbL film involved the alternated adsorption from solution of both polyelectrolytes, namely, the substrate was: (1) immersed in the PAH solution, for 3 min; (2) washed with ultrapure water; (3) immersed in the PAZO solution, for 3 min; (4) washed with ultrapure water. In order to obtain a film with n bilayers, denoted as (PAH/PAZO)$_n$, the aforementioned sequence is repeated n times.

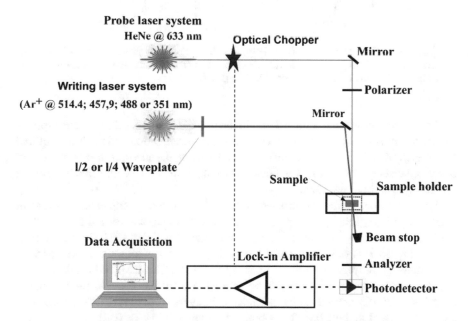

Fig. 11.12 Experimental setup layout used to characterize the dynamics of creation, decay and deletion of birefringence in azobenzene containing films

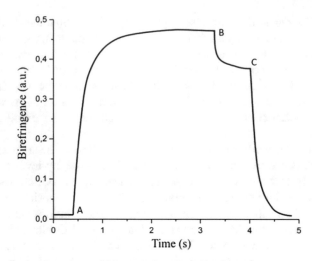

Fig. 11.13 Illustrative plot of the creation, relaxation and erase processes of birefringence as a function of time. A—Starting of irradiation with linearly polarized light. B—Interruption of irradiation. C—Irradiation with circular polarized light

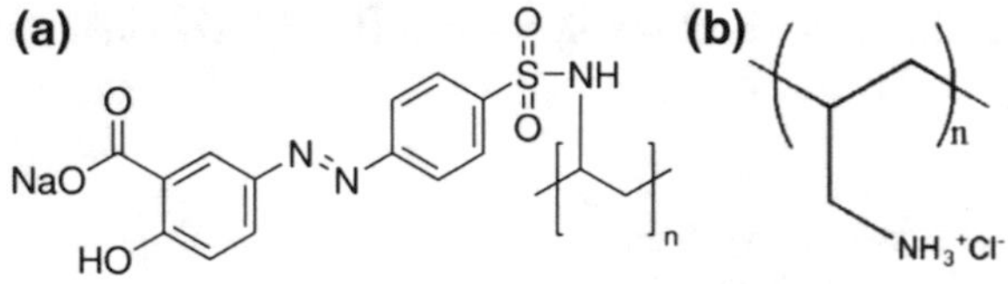

Fig. 11.14 Chemical structures of: **a** Poly{1-(4-(3-carboxy-4-hydroxy phenylazo) benzene sulfonamido)-1,2-ethanediyl, sodium salt} (PAZO) , a main chain azobenzene containing polyelectrolyte, and **b** Poly (allylamine hydrochloride) (PAH)

The chosen immersion time of 3 min in the polyelectrolytes solutions corresponds to a compromise between optimized time for films preparation and the time required to achieve 80% of adsorbed amount per layer, based on the adsorption kinetics studies described elsewhere [34, 43]. The thickness of the films can be controlled by keeping the adsorption times constant.

The ultraviolet-visible absorbance spectra for different number of bilayers of PAH/PAZO LBL films are depicted in Fig. 11.15a. The absorbance peak centred near 360 nm is related to the $\pi - \pi^*$ chromophore transition [37]. By the analysis of Fig. 11.15b it is observed that the absorbance at the maximum increases with the number of bilayers indicating a linear film growth. The adsorbed PAZO amount per unit of area and per bilayer can be estimated from the slope of the maximum absorbance at 363 nm versus the number of bilayers in the PAH/PAZO LbL film, using the Beer-Lambert law. The obtained value is 0.014 ± 0.001 mg m^{-2}, and the PAZO absorption coefficient, at 363 nm, $\varepsilon_{363nm} = 4.30 \pm 0.07$ m^{2}g^{-1} [35]. The

Fig. 11.15 **a** Absorption spectra of PAH/PAZO LbL films with different number of bilayers. **b** Maximum absorbance (363 nm) versus the number of bilayers in the PAH/PAZO LbL film

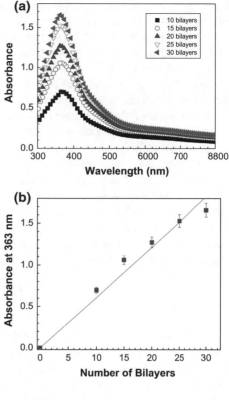

Fig. 11.16 Birefringence creation and relaxation kinetics curve at room temperature obtained for (PAH/PAZO)$_{30}$ LbL films with polarized visible light

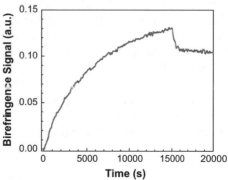

adsorbed PAZO amount achievable for 30 bilayers film is 0.42 ± 0.03 mg m^{-2}, which is a high value in comparison with films prepared onto glass substrates [38].

In Fig. 11.16 A is presented a typical birefringence creation and relaxation kinetics curve obtained for 30 bilayers PAH/PAZO LbL films.

As already mentioned the birefringence creation curves correspond to the increase of transmitted light signal when the writing beam source is turned on, whereas the birefringence relaxation curves are acquired immediately after the light to be turned off and correspond to a decrease in the transmission signal. A 650 nm 0.9 mW laser diode module was employed to measure the photoinduced birefringence. The sample was placed between crossed polarisers, in what concerns to the probe beam, with polarization set at 45° with respect to the writing beam polarization. An optical chopper was used to modulate the probe beam and the birefringence signal was measured with a photodetector, through a lock-in amplifier.

The analysis of birefringence creation kinetics curves can be carried out by fitting the experimental data to a bi-exponential function, containing two distinct processes in agreement with literature, see [36] and references therein. Generally, the fast process is assigned to *trans→cis→trans* photo-isomerization processes contributing to the birefringence, which depends on the available free local volume and on interactions between chromophores and PAH. The slower process is ascribed to the main chain mobility, which relies on chain size and interactions between both polyelectrolytes. However, in this study, the birefringence creation, $I_{writing}$, was seen to be very slow and therefore the experimental data can be adequately fitted only with a single exponential function as:

$$I_{writing} = I_{w0}\left(1 - \exp\left(-\frac{t}{\tau_w}\right)\right) \tag{11.7}$$

where, I_{w0} is the pre-exponential factor that represents the magnitude of the process and τ_w is the characteristic time constant. In this case, the characteristic time corresponds to a value of 5180 ± 20 s, which is a much longer time in comparison with the value of 1700 ± 40 s obtained in PAH/PAZO films prepared onto glass substrates, using as writing beam the 514 nm beamline of a tuneable Ar+ laser [38]. This means that the incidence of light is likely to increase the sample temperature, avoiding in this way the creation of birefringence. It can be also detected that the induced birefringence is proportional to the amount of azo-groups present in the films, i.e., of the number of PAZO layers.

The obtained decay curve can be fitted with two exponential Debye like processes, being the one with short characteristic associated with dipole disorientation and the one with longer characteristic time, the long-term relaxation, related to disorientation arising from the movement of polymer chains, as follows:

$$I_{relaxation} = I_{r1} \exp\left(\frac{t}{\tau_{r1}}\right) + I_{r2} \exp\left(-\frac{t}{\tau_{r2}}\right) \tag{11.8}$$

where I_{r1} and I_{r2} are the pre-exponential factors for the normalized birefringence intensity, τ_{r1} and τ_{r2} are the characteristic time constants of the processes. For the earliest moments of relaxation process, the obtained characteristic time is a value of 1970 ± 30 s, while the second characteristic time is of $(0.63 \pm 0.03) \times 10^6$ s. These values are in good agreement with the birefringence decay characteristic times of PAZO [36].

11.6 Conclusions

This work reviews a new concept towards energy harvesting from light based on dielectrics media containing highly polarisable chromophores, which are photosensitive—The so called photo-electret. Materials techniques for device production were presented. Particularly, two molecular systems were analysed, the multifunctional diarylethenes and azobenzenes derivatives. Blends of the photochromic material DTE-CN revealed to produce a displacement current density to an external circuit when irradiated with UV light at 375 nm. It was further revealed that birefringence can be successfully induced in PAH/PAZO LbL films by irradiation with light with a spectrum similar to that of sun light. In ideal conditions an efficiency of 16% can be expected, which is a fairly good starting point for further investigations and making the concept of photo-electret based harvesting devices promising.

Acknowledgements The authors acknowledge the financial support from FEDER, through Programa Operacional Factores de Competitividade—COMPETE and Fundação para a Ciência e a Tecnologia—FCT, for the projects UID/FIS/00068/2013, POCTI/FAT/47529/2002 and PTDC/FIS-NAN/0909/2014.

References

1. C.A. Hill, M.C. Such, D. Chen, J. Gonzalez, W.M. Grady, Battery energy storage for enabling integration of distributed solar power generation. IEEE Trans. Smart Grid **3**(2), 850–857 (2012)
2. N.S. Lewis, Towards cost-effective solar energy use. Science **315**(5813), 798–801 (2007)
3. K. Wang, H. Wu, Y. Meng, Z. Wei, Conducting polymer nanowire arrays for high performance supercapacitors. Small **10**(1), 14–31 (2014)
4. C. Hu, L. Song, Z. Zhang, N. Chen, Tailored graphene systems for unconventional applications in energy conversion and storage devices. Energy Environ. Sci. **8**, 31–54 (2014)
5. H. Wang, Y. Liang, T. Mirfakhrai, Z. Chen, H.S. Casalongue, H. Dai, Advanced asymmetrical supercapacitors based on graphene hybrid materials. J. Nano Res. **4**(8), 729–736 (2011)
6. G.M. Sessler, (ed.), *Electrets-Topics in Applied Physics*, vol. 33 (Springer, Berlin Heidelberg, 1980, 1987). ISBN: 978-3-540-17335-9 (Print), 978-3-540-70750-9 (Online)
7. O. Heaviside, Electromagnetic induction and its propagation. Electrization and electrification. Natural electrets. Electrician 230–231 (1885)
8. G.M. Sessler, J.E. West, Self-biased condenser microphone with high capacitance. J. Acoust. Soc. Am. **34**, 1787 (1962)
9. M. Eguchi, On the permanent electret. Philos. Mag. **48**, 178–192 (1925)
10. E. Fukada, Piezoelectricity in polymers and biological materials. Ultrasonics **6**, 229 (1968)
11. M. Raposo, P.A. Ribeiro, J.N. Marat-Mendes, Interaction of corona discharge species with teflon-FEP (Fluoroethylenepropylene) foils. Ferroelectrics **134**(1–4), 235–240 (1992)
12. H. Kawai, The piezoelectricity of poly (vinylidene fluoride). Jpn. J. Appl. Phys. **8**(7), 975 (1969)
13. A.J. Lovinger, Ferroelectric polymers. Science **220**(4602), 1115–1121 (1983)
14. T. Furukawa, Ferroelectric properties of vinylidene fluoride copolymers. Phase Trans. **18**, 143–211 (1989)
15. J.A. Giacometti, P.A. Ribeiro, M. Raposo, J.N. Marat-Mendes, J.S.C. Campos, A.S. De Reggi, Study of poling behavior of biaxially stretched poly (vinylidene fluoride) films using the constant-current corona triode. J. Appl. Phys. **78**(9), 5597–5603 (1995)

16. F.I. Mopsik, M.G. Broadhurst, Molecular electrets. J. Appl. Phys. **46**(10), 4204–4208 (1975)
17. X. Zhang, G.M. Sessler, Y. Wang, Fluoroethylenepropylene ferroelectret films with cross-tunnel structure for piezoelectric transducers and micro energy harversters. J. Appl. Phys. **116**(074109), 1–8 (2014)
18. X. Zhang, W. Liming, G.M. Sessler, Energy harvesting from vibration with cross-linked polypropylene piezoelectrets. AIP Adv. **5**(077185), 1–10 (2015)
19. P. Podrom, S.G.M. Hillenbrand, J. Bos, T. Melz, Vibration-based energy harvesting with stacked piezoelectrets. Appl. Phys. Lett. **104**(172901), 1–5 (2014)
20. X. Zhang, P. Podrom, L. Wu, G.M. Sessler, Vibration-based energy harvesting with piezoelectrets having high d$_{31}$ activity. Appl. Phys. Lett. **108**(193903), 1–4 (2016)
21. A. Cuadras, M. Gasulla, V. Ferrari, Thermal energy harvesting through pyroelectricity. Sens. Actuators A **158**, 132–139 (2010)
22. R. Castagna, M. Garbugli, A. Bianco, S. Perissinotto, G. Pariani, C. Bertarelli, G. Lanzani, Photochromic electret: a new tool for light energy harvesting. J. Phys. Chem. Lett. **3**, 51–57 (2012)
23. J.H. Kim, K.M. Hong, H.S. Na, Y.K. Han, Polymer thin films containing azo dye for rewritable media. Japanese J. Appl. Phys. **40**, 1585–1587 (2001)
24. A. Natanshon, P. Rochon, J. Gosselin, S. Xie, Azo polymers for reversible optical storage 1. Poly[4'-[[2-(acryloyloxy)ethy]ethylamino]4-nitroazobenzene]. Macromolecules **25**, 2268–2273 (1992)
25. X. Meng, A. Natanshon, P. Rochon, Azo polymers for reversible optical storage. 11 poly{4,4'-(1-methylethylidene) bisphenylene 3-[4-(4-nitrophenylazo)phenyl]-3-aza-pentanedioate}. J. Polym. Sci. Part B Polym. Phys. **34**, 1461–1466 (1996)
26. Z. Sekkat, J. Wood, W. Knoll, Reorientation mechanism of azobenzenes within the trans-cis photoisomerization. J. Phys. Chem. **99**, 17226–17234 (1995)
27. G.S. Hartley, The cis-form of azobenzene. Nature **140**, 281 (1937)
28. K.G. Yager, C.J Barret, Azobenzene polymers for photonic application. 1–35 (2009). Wiley
29. A. Monteiro, Estudo da dinâmica da criação e da relaxação de birrefrigência fotoinduzida em filmes automontados de PAH/PAZO, MSc thesis, Universidade Nova de Lisboa, Lisboa, Portugal (2013)
30. C. Cojocariu, P. Rochon, Light-induced motions in azobenzene containing polymers. Pure Appl. Chem. **76**, 1479–1497 (2004)
31. H. Rau, Photoisomerization of sterically hinderes azobenzes. J. Photochem. Photobiol. **42**, 321–327 (1988)
32. P. Farinha, S. Sério, P.A. Ribeiro, M. Raposo, Birefringence creation by solar light—a new approach to the development of solar cells with azobenzene materials, in *Proceedings of the 4th International Conference on Photonics, Optics and Laser Technology (PHOTOPTICS 2016)*, pp. 367–370. ISBN: 978-989-758-174-8
33. O.N. Oliveira Jr., M. Raposo, A. Dhanabalan, Langmuir-Blodgett (LB) and self-assembled (SA) polymeric films, in *Handbook of Surfaces and Interfaces of Materials*, vol. 4, Chapter 1, ed. by H.S. Nalwa (Academic Press, New York), pp. 1–63
34. Q. Ferreira, P.J. Gomes, M. Raposo, J.A. Giacometti, O.N. Oliveira Jr., P.A. Ribeiro, Influence of ionic interactions on the photoinduced birefringence of poly [1-[4-(3-Carboxy-4 Hydroxyphenylazo) benzene sulfonamido]-1,2-ethanediyl, sodium salt] films. J. Nanosci. Nanotechnol. **7**, 2659–2666 (2007)
35. Q. Ferreira, P.J. Gomes, M.J.P. Maneira, P.A. Ribeiro, M. Raposo, Mechanisms of adsorption of an azo-polyelectrolyte onto layer-by-layer films. Sens. Actuators B **126**, 311–317 (2007)
36. Q. Ferreira, P.A. Ribeiro, O.N. Oliveira Jr., M. Raposo, Long-term stability at high temperatures for birefringence in PAZO/PAH layer-by-layer films. ACS Appl. Mater. Interfaces **4**, 1470–1477 (2012)

37. Q. Ferreira, P.J. Gomes, P.A. Ribeiro, N.C. Jones, S.V. Hoffmann, N.J. Mason, O.N. Oliveira Jr., M. Raposo, Determination of degree of ionization of poly (allylamine hydrochloride) (PAH) and poly[1-[4-(3-carboxy-4 hydroxyphenylazo) benzene sulfonamido]-1,2-ethanediyl, sodium salt] (PAZO) in layer-by-layer films using vacuum photoabsorption spectroscopy. Langmuir **29**(1), 448–455 (2013)
38. A.R. Monteiro-Timóteo, J.H.F. Ribeiro, P.A. Ribeiro, M. Raposo, Dynamics of creation photoinduced birefringence on (PAH/PAZO) n layer-by-layer films: analysis of consecutive cycles. Opt. Mater. **51**, 18–23 (2016)
39. J.A. Giacometti, J.S.C. Campos, Constant current corona triode with grid voltage control—application to polymer charging. Rev. Sci. Instrum. **61**(3), 1143–1150 (1990)
40. B. Gross, R. Hessel, Electron-emission from electron-irradiated dielectrics. IEEE Trans. Electr. Insul. **26**(1), 18–25 (1991)
41. R. Piron, E. Toussaere, D. Josse, S. Brasselet, J. Zyss, Towards non-linear photonics in all-optically poled polymer microcavities. Synth. Met. **115**, 109–119 (2000)
42. C. Madruga, P. Alliprandini, M.M. Andrade, M. Goncalves, M. Raposo, P.A. Ribeiro, Birefringence dynamics of poly{1-[4-(3-carboxy-4 hydroxyphenylazo) benzene-sulfonamido-1,2-ethanediyl, sodium salt} cast films. Thin Solid Films **519**(22), 8191–8196 (2011)
43. J.J. Ramsden, Y. Lvov, G. Decher, Determination of optical constants of molecular films assembled via alternate polyion adsorption. Thin Solid Films **254**, 246–251 (1995)

Author Index

© Springer Nature Switzerland AG 2018
P. A. Ribeiro and M. Raposo (eds.), *Optics, Photonics and Laser Technology*, Springer
Series in Optical Sciences 218, https://doi.org/10.1007/978-3-319-98548-0

Subject Index

© Springer Nature Switzerland AG 2018   245
P. A. Ribeiro and M. Raposo (eds.), *Optics, Photonics and Laser Technology*, Springer
Series in Optical Sciences 218, https://doi.org/10.1007/978-3-319-98548-0

Printed in the United States
By Bookmasters